International Political Economy Series

Series Editor
Timothy M. Shaw
University of Massachusetts
Boston, MA, USA

and

University of London
London, UK

The global political economy is in flux as a series of cumulative crises impacts its organization and governance. The IPE series has tracked its development in both analysis and structure over the last three decades. It has always had a concentration on the global South. Now the South increasingly challenges the North as the centre of development, also reflected in a growing number of submissions and publications on indebted Eurozone economies in Southern Europe. An indispensable resource for scholars and researchers, the series examines a variety of capitalisms and connections by focusing on emerging economies, companies and sectors, debates and policies. It informs diverse policy communities as the established trans-Atlantic North declines and 'the rest', especially the BRICS, rise.

More information about this series at
http://www.springer.com/series/13996

Johannes Knierzinger

Bauxite Mining in Africa

Transnational Corporate Governance and Development

palgrave
macmillan

Johannes Knierzinger
Leipzig University
Toulouse, France

International Political Economy Series
ISBN 978-3-319-52705-5 ISBN 978-3-319-52706-2 (eBook)
DOI 10.1007/978-3-319-52706-2

Library of Congress Control Number: 2017943492

Printed on acid-free paper

This Palgrave Macmillan imprint is published by Springer Nature
The registered company is Springer International Publishing AG
The registered company address is: Gewerbestrasse 11, 6330 Cham, Switzerland

For Lulu and Claire.

Acknowledgements

This work is the outcome of a quite laborious combination of the literature on the political and social history of Guinea with primarily business-oriented information on the global aluminum industry, including 5 months of research in Guinea's bauxite towns and the Guinean capital, Conakry. As you can imagine, I particularly enjoyed the second part of this study: Guinea is a thrilling and pulsating country and I met many broad-minded and enquiring people in places, where I would not have expected such people to be. However, even under the formally democratic regime of Alpha Condé since 2011, the freedom of expression is not always guaranteed, particularly when foreign greenhorns have too many questions about the mining industry. Above all, I thus would like to sincerely thank those friends and informants who decided to talk openly about the living and working conditions in Guinea, thereby sometimes endangering their good relations with the establishment and even their jobs. In order to protect those who could be negatively affected, I will not mention any Guinean names here, but, nonetheless, I thank all my hosts and informants for the countless hours of support, the nice meals, the entertaining evenings, and most of all for all the precious advices and insights. As a matter of course, this work would neither have been possible without the constant advice and the guidance by my colleagues, professors, and friends in Germany, France, Austria, and the US. Among them, I would like to especially thank Emmanuel Grégoire, Ulf Engel, Géraud Magrin, Matthias Middell, Melanie Pichler, Karin Fischer, Walter Schicho, Micha Fiedlschuster, Johannes Frische, David Mayer,

Helena Flam, Ute Rietdorf, Johanna Rau, and Jess Auerbach. Last but not least, I am also in great debt to my ever-growing extended family: thank you for making this odyssey possible.

Contents

Acronyms

ACG	Alumina Compagnie de Guinée
AFD	Agence Française de Développement
AIAG	Aluminium Industrie AG
AIDS	Acquired Immune Deficiency Syndrome
AMC	Alliance Mining Commodities
ANAIM	Agence Nationale d'Aménagement des Infrastructures Minières
AOF	Afrique-Occidentale Française
AOS	Aluminium Oxid Stade GmbH
ARSYF	Association des Ressortissants et Sympathisants de Fria
ASI	Aluminium Stewardship Initiative
ASV	Årdal og Sunndal Verk
BIT	Bilateral Investment Treaty
BRICS	Brazil, Russia, India, China and South Africa
CATIC	China Aeronautic Technology and Industry Corporation
CBG	Compagnie des Bauxites de Guinée
CBK	Compagnie des Bauxites de Kindia
CC	Commodity Chains
CEO	Chief Executive Officer
CFA	Coopération Financière en Afrique
CGG	Compagnie Guinéenne de Génie
CMRN	Comité Militaire de Redressement National
CNDD	Conseil National pour la Démocratie et le Développement
CNSS	Caisse Nationale de Sécurité Sociale
CNTG	Confédération Nationale des Travailleurs de Guinée
COBAD	Compagnie de Bauxite de Dian-Dian
CPD	Conseil Préfectoral de Développement

CPI	China Power Investment
CSA	Centre de Santé Amelioré
CSR	Corporate Social Responsibility
CU	Commune Urbaine
DFG	Deutsche Forschungsgemeinschaft
DGA	Directeur Général Adjoint
DRC	Democratic Republic of the Congo
DSPJ	Directeur Sous-préfectoral de la Jeunesse
DUBAL	Dubai Aluminium
EDF	Electricité de France
EDG	Electricité de Guinée
EGA	Emirates Global Aluminium
EGB	Entretien Général du Bâtiment
EIB	European Investment Bank
EITI	Extractive Industries Transparency Initiative
EMF	Entretien et Maintenance Friguia
EMG	Entretien et Maintenance en Guinée
FOM	France d'Outre Mer
GAC	Guinea Alumina Corporation
GAZ	Gorkovsky Avtomobilny Zavod
GCC	Global Commodity Chains
GDP	Gross Domestic Product
GIZ	Deutsche Gesellschaft für Internationale Zusammenarbeit
GNF	Franc guinéen
GNP	Gross National Product
GVC	Gobal Value Chains
HIPC	Heavily Indebted Poor Countries
HIV	Human Immunodeficiency Virus
IBA	International Bauxite Association
IBRD	International Bank for Reconstruction and Development
IDC	Industrial Development Development
IFC	International Finance Corporation
IHA	Institut pour l'Histoire de l'Aluminium
IMD	International Mining Company
IMF	International Monetary Fund
ISO	International Organization for Standardization
LME	London Metal Exchange
MARG	Mission d'aménagement de Guinée
Mt	Megaton
MW	Megawatt
NGO	Non-governmental organization
NRGI	Natural Resource Governance Institute

OBK	Office des Bauxites de Kindia
OEC	Observatory of Economic Complexity
OECD	Organisation for Economic Co-operation and Development
OFAB	Office d'Aménagement de Boké
OPEC	Organization of the Petroleum Exporting Countries
PACV	Programme d'appui Aux Communautés Villageoises
PAM	Provision Alimentaire Mensuelle
PDD	Plan de développement de district
PDL	Plan de déplacement Local
PUK	Pechiney Ugine Kuhlman
PWYP	Publish What You Pay
RWI	Revenue Watch Institute
SAG	Société Ashanti de Guinée
SAP	Structural Adjustment Program
SAREPA	Société africaine de recherche et d'études pour l'aluminium
SBDT	Société des Bauxites de Dabola—Tougué
SBK	Société des bauxites de Kindia
SEMF	Société électrométallurgique française
SENAP	Service national des points d'eau
SMD	Société Minière de Dinguiraye
STABEX	Système de Stabilisation des Recettes d'Exportation
SYSMIN	System of Stabilization of Export Earnings from Mining Products
TCA	Taxe sur le chiffre d'affaires
TNC	Transnational Corporation
USADF	US African Development Foundation
USD	US-Dollar
USGS	United States Geological Survey
VAW	Vereinigte Aluminium-Werke
VIAG	Vereinigte Industrieunternehmungen AG
WTO	World Trade Organization
WWII	Second World War

List of Figures

List of Tables

CHAPTER 1

Introduction

From the 1930s, a few African bauxite mines—in Guinea, Ghana, Sierra Leone, Mozambique, and later also in Tanzania—became part of a worldwide production network of mines, refineries, smelters, and metal-working factories, all of which were controlled by a few interconnected companies. This book focuses on the consequences of this inclusion into the global aluminum chain, which results in cars, airplanes, buildings, electronic devices, cans, and other consumer goods. Around the decade of the 1960s, when most African countries became independent, they did not possess the technical and political knowledge necessary to build up whole states. Sékou Touré, Guinea's head of state from 1958 until his death in 1984, only counted 12 Guinean citizens with a university degree at the time of independence (Touré 1967: 37) and Guinea's population was almost entirely illiterate at this time (Shundeyev 1974: 36). Ghana counted 29 law students (studying in the UK) in 1948 (Rathbone 1999: 52ff).

At the same time, both governments were dependent on the export of a very limited set of raw materials. The revenues of the Guinean government came almost entirely from bauxite mines and a connected refinery (Larrue 1997: 92; Campbell and Clapp 1995: 427), which were conjointly managed by European, North American, and Russian aluminum enterprises. Besides their influence on national politics, these companies created and de facto ruled over company towns with several hundred thousand inhabitants. They also possessed the majority of the bauxite mines in the rest of the world and thereby controlled a great deal of the

J. Knierzinger, *Bauxite Mining in Africa*, International Political Economy Series, DOI 10.1007/978-3-319-52706-2_1

strings in Africa's bauxite producing countries that had previously been directly pulled by the colonialists. The political consequences of this new power configuration, and the explication of how this "rule" was different from former systems of domination, such as colonialism, are the central topic of this book. The simplified picture of almighty Northern companies controlling a poor Southern "developing country" is complicated by the fact that these companies were strongly tied to nation states, interconnected with each other, and certainly also connected to the African elite.

Serious challenges in terms of actually putting down my findings in writing included both the long time frame—from independence until today—as well as the different scales in play. The mining companies exercise a controlling influence on their company towns, interact with governments, and are confronted with political and economic developments around the globe, including African social movements. The imperfect solution to this problem of display was to start with a brief description of the global aluminum production and its entanglements with the mentioned African countries in this chapter. In Chap. 2, I will present the conceptual framework of this study. In Chap. 3, the story will zoom back to the situation in the company towns.

Besides the scarce existing academic literature on bauxite mining in Africa, the findings of this book are mainly based upon field research in Guinea (February 2012 and January to April 2014), business literature and articles mainly derived from the internet, as well as research in the archive of Pechiney (now Rio Tinto Alcan) at the *Institut pour l'Histoire de l'Aluminium* (IHA) in Paris. The most valuable basis of this book was the work of the political economist Bonnie Campbell from the University of Quebec. Apart from her work, most of the literature on African bauxite mining has been written by former politicians, administrators, and managers.

This concerns accounts of the former Guinean minister for mining, Ibrahima Soumah (Soumah 2008), whose opus magnum has lately even been translated into Chinese (Interview with Engineer CBG 6/02/2014), as well as most of the literature on the company towns of Fria and Edéa (Pauthier 2002: 4–5; Grinberg 1994). Soumah was in power from 2000 to 2002, during a time when Guinean politics was in a process of disintegration and did not have much credibility in the eyes of my interviewees (Interview with Former executive Rusal 3/04/2014). Thanks to the IHA, a considerable amount of the literature on the history of Pechiney in Africa has emerged, although most of it has been written by either veterans of Pechiney or French government officials themselves (see Larrue 1991; Laparra 1995) or was almost exclusively based upon accounts of the company. For instance, Jacques Larrue was *Administrateur de la France d'outre-mer, Chef de cironscription* and later *Inspecteur du travail* and thus was part of the (post-)colonial enterprise before he wrote his book on Fria (Laparra 1995: 422).[1] Maurice Laparra was director of the smelter in Edéa before he wrote a book on this topic and later became *directeur général* of Pechiney and president of the *Institut pour l'Histoire de l'Aluminium* (Thaure 2007: 10; see also Grinberg 1994; Institut pour l'Histoire de l'Aluminium 2015). As the entire public archives of the regime under Sékou Touré have been destroyed and the remaining documents often remain in sealed cases under the desks of government officials, a good part of Guinean history remains to be written. While my own contribution to the history of Guinean bauxite mining before 1980 remained punctual, developments in the bauxite towns since the 2000s have received much attention in this book, in particular with regard to the eventful recent history of Fria.

Most of the time that I spent in Guinea was dedicated to carrying out more than 150 semi-structured interviews, which were spread equally across the four Guinean bauxite towns—Sangaredi, Kamsar, Kindia/Débélé, and Fria—and the capital Conakry. This meant that I was constantly moving from one place to another during 5 months of my research. During my stay in Sangaredi in 2012, I was kindly housed in the city center by the bauxite company itself, and received a guided tour of the mines. There, I mostly had contacts with senior officers and expatriates. In 2014, I stayed mostly with the families of workers.

1.1 The Global Production Network of Aluminum

In order to treat the main question of the sociopolitical repercussions of African bauxite mining, it is first necessary to explain how and where aluminum and its raw material bauxite are produced and which companies are involved in this process. Primary aluminum, which has to be processed and blended before it can be used, is extracted in a very energy-intensive process from aluminum oxide (also called alumina), which in turn is refined out of bauxite, the basic material of the aluminum production chain (see Fig. 1.1).

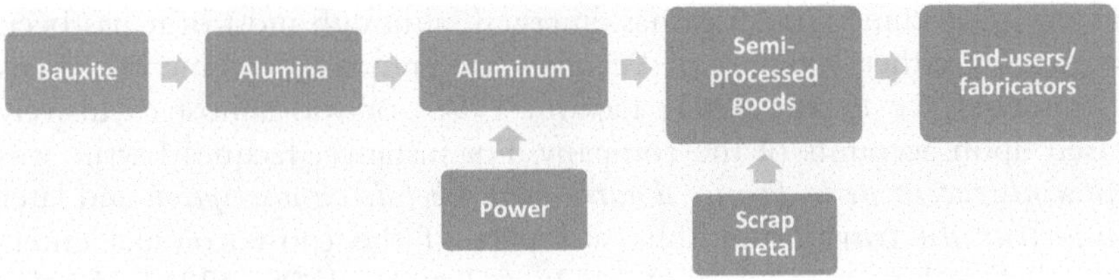

Fig. 1.1 Production chain of aluminum

From a global historical perspective, two "aluminum reigns" can be distinguished since WWII: first, US domination from 1943 until the oil crisis and second, China's impressive rise from controlling about 10% of the market in 2000 to over 50% today.[2] The end of the German domination of the aluminum sector in 1943 coincides with the steep rise of US production, which was an essential condition for the US victory in WWII, primarily because airplanes almost entirely consist of aluminum. This commanding position—marked by control of more than two-thirds of world production of aluminum together with the smelters in Canada, which were owned by US producers—was subsequently challenged by Russia, which also expanded strongly during WWII,[3] and the return of European producers, such as France, Norway, Switzerland, Germany, and many other small European countries, as well as Japan. Production of aluminum by the latter, however, was stopped particularly violently by the oil crisis.[4] The majority of European producers were able to maintain their shares of global production until the dissolution of the Soviet Union and the start of the Chinese ascent and—to a lesser degree—other Asian and African producers (mainly India and South Africa). Whereas the steep rise of aluminum production in the US was a direct consequence of war mobilization, China primarily needs the material for its infrastructural projects and to fulfill the wishes of worldwide consumers (Fig. 1.2).[5]

Absolute aluminum production increased significantly in the twentieth century. During the first production peak in 1943, the world produced 1400 Mt of aluminum, compared to the next major peak in 1979 at 16,000 Mt and the current production of about 50,000 Mt. Measured in terms of absolute all-time production, China already overtook Russia in the mid-2000s and is expected to surpass the US in the next years to come, despite only starting its ascent in the late 1990s. From the

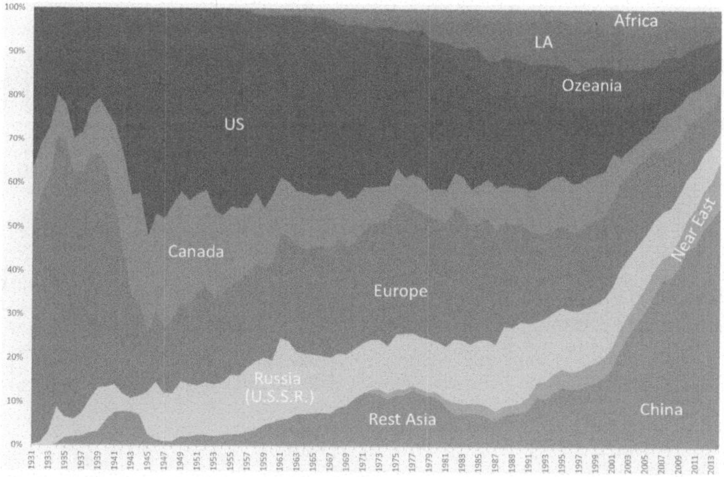

Fig. 1.2 Worldwide primary aluminum production by country/region in percent, 1931–2013. *Source* USGS (2016)

start of production until the present, annual aluminum production has declined six times, each time reflecting global turning points: from 1943 to 1946, 1974 to 1975, 1979 to 1982, slightly from 1984 to 1986, and from 1993 to 1994, and again considerably from 2008 to 2009 (USGS 2016). Both the downturns in the 1970s and the recent recession led to significant reductions of smelting capacity in Europe, of about 25% again during the recent crisis. The reasons remained the same as in the 1970s: escalating power tariffs, lack of local ores, higher taxation, and higher ecological standards (Shoeb 2012) (Fig. 1.3).

At first glance, the aluminum sector appears to be less concentrated in the 2000s compared with the 1970s in terms of company shares. Similar to the so-called *Seven Sisters* of the global oil business, the global aluminum chain was controlled by *Six Sisters* until the 1970s, namely, Alcoa, Reynolds, Kaiser (all from the US), Alcan (Canada), Alusuisse (Switzerland), and Pechiney (France). While production in the US dwindled to less than 5% of the current worldwide production, the remaining major US producer and the only survivor of the Six Sisters, Alcoa, still retain a share of about 10% of worldwide production and operates in 30 countries across the world (Alcoa 2015). This high percentage of foreign assets is also discernible in the case of the Norwegian Norsk Hydro,

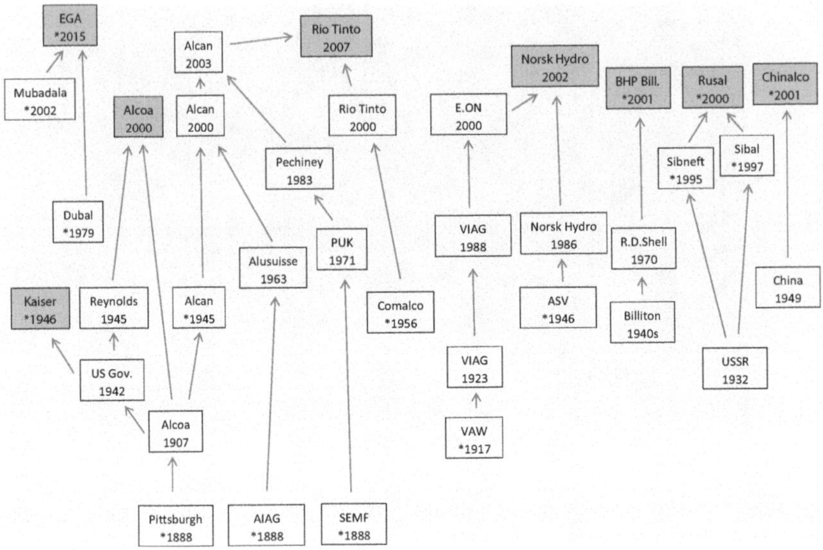

Fig. 1.3 History of major corporations of the global aluminum industry, 1888–2016

whereby the Canadian smelters—still producing about 7% of worldwide production—belong to the production network of the multinational Rio Tinto since 2007, which has its head offices in London and Melbourne. Together with Rusal, Chinalco, and Alcoa, Rio Tinto is currently one of the four largest aluminum companies, each producing about one-tenth of worldwide primary aluminum. Despite some Western companies with minor stakes (such as Alcoa with an 8% stake in Chinalco) and vice versa (Chinalco has stakes in Australia, Peru, Vietnam, and acquired a 9% stake of Rio Tinto),[6] Chinese production is mostly centered on Chinese demand and steered by Chinese companies. Among the world's ten largest aluminum companies, five are Chinese (Rusal 2014a), with the majority of these companies being state-owned. The ascent of China thereby led to another concentration of the sector after a long period of deconcentration since WWII. The Six Sisters and the Russian government controlled about three-quarters of aluminum production in the early 1960s and two-thirds in the mid-1970s (USGS 2016). If we consider China as one state-owned producer with a current share of about 50% and add the four next largest company producers, the current concentration is even higher than in the 1960s (Fig 1.4).

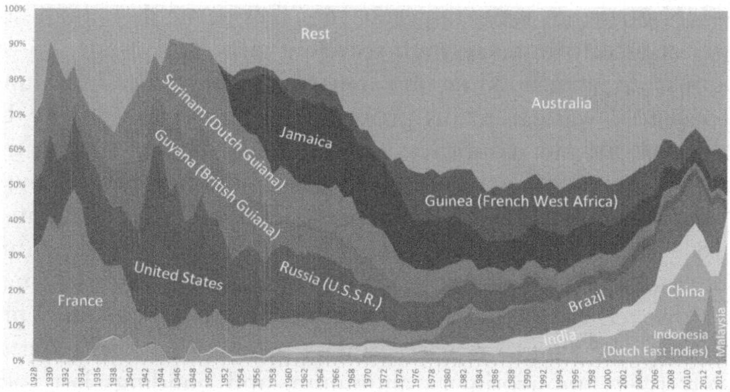

Fig. 1.4 Worldwide bauxite production per country in percent, 1928–2012. (*Source* USGS 2016)

In worldwide bauxite production, five peaks in *shares* can be identified: first, the domination of French bauxite in the 1930s, which was not matched by an equally dominant French aluminum production, because the bauxite had been exported to other European countries (from 1940 to 1945 mainly to Germany; see Bunker and Ciccantell 1995: 60; Wiederstein 1994: 12); second, the US expansion after its entry into the war in the 1940s; third, a steep rise in Caribbean production in the early 1950s, which was mainly exported to the US; fourth, the rise of Australian production as a reaction to a Caribbean-led revolt of third-world producers since the 1960s, which was also followed by the expansion of Guinean and Brazilian production since the mid-1970s; and fifth, the rise of Asian production. Each of these peaks amounted to almost half of worldwide production (France—US—Caribbean countries—Australia), with the exception of the last expansion, which nonetheless led to a dominating position for newly industrialized and resource-rich countries (Brazil, China, India, and Indonesia).

1.1.1 The Social and Ecological Consequences of Aluminum Production

Aluminum smelting is one of the most energy-intensive industrial processes in existence. The annual hydroelectric power used in worldwide aluminum production nearly equals the annual power demand of Australia (Riverton 2014: 35), or of all Africans living south of the

Sahara (UN 2015; cf. Knierzinger 2016). This particularly high demand for power also contributes to high recycling rates. In Europe, 85–95% of aluminum scrap from building and transport is recycled. Only aluminum from household waste remains problematic, particularly in the United States. Lower income countries seem to have higher recycling rates; for instance, Brazil has a recycling rate of aluminum cans of over 90% (cf. Brinkmann 2004: 15–17; Hildebrand 2007: 47; Hütz-Adams et al. 2014: 18). In West Africa, the artisanal smelting of secondary aluminum is widespread and has been practiced since the 1940s (Romainville 2009). However, for technical and economic reasons, secondary production will never be able to replace primary production. Packaging almost exclusively relies upon primary aluminum, and buildings last too long to contribute sufficiently to recycling as long as world demand is growing (cf. Brinkmann 2004: 16; Track Record 2010: 14). The huge energy demand for primary aluminum smelting is thereby especially problematic in the case of the packaging industry. Even if recycling of aluminum cans was at 100%, they would still consume about ten times more energy than returnable glass bottles (Bund der Energieverbraucher 2015).

Aluminum smelting also leads to pollution through fluoride emissions. Wiederstein reported that millions of hectares of forests in Russia have been destroyed owing to such emissions (Wiederstein 1994: 43). The needed energy either pollutes the air directly—if it is produced with coal and gas—or leads to the construction of huge dams with comparable negative impacts, such as the destruction of forests, forced migration, and high greenhouse emissions. Recent studies suggest that large dams sometimes produce much more greenhouse gas than power plants running on fossil fuels (Graham-Rowe 2005). Due to its low weight and high recyclability, the use of aluminum in transportation is currently considered environmentally acceptable (e.g., Kerkow et al. 2012: 26), although this perception could change quickly in the next years due to nanotechnological innovations that enable the replacement of aluminum by steel in aircraft and car production (cf. The Economist 2015). The excessive use of aluminum in packaging has historical reasons (including overproduction in the US after WWII) and could easily be reversed by the (re-)introduction of various ecologically more sustainable substitutes. The current debates on the health risks of the use of aluminum in the food and cosmetics industry (see Bundesministerium für Gesundheit 2014)[7] could also have an impact on these ecological discussions.

Bauxite mining leads to the expropriation of land, forced migration, deforestation, water shortages, land conflicts, and respirational diseases.

Aside from the work of Bonnie Campbell—who has mainly focused on politico-economic aspects of bauxite mining—these consequences have not received academic attention to date. Environmental and social impact studies of the mining companies have begun during the recent mining boom, albeit only covering a small percentage of the mining areas by the time of writing (Interview with Anthropologist 4/02/2014; Sociologist 22/02/2014; Executive CBG 23/02/2014). In addition to these consequences for the local population, the working conditions of those who labor on the mines are harsh. They suffer from respirational diseases, deplore low salaries (particularly for workers on contractual basis), and especially low pensions.

The commission of the Fria alumina refinery led to about 30 years of entirely uncontrolled pollution by the so-called red mud and other by-products. This destroyed the livelihood of thousands of farmers and fishermen and led to the death of numerous children who fell into the red mud dumps. The last of these cases was reported in 2008 when both the government and the companies have engaged in some efforts to stop these industrial wastes from contaminating rivers and groundwater, although such measures have remained inefficient and insufficient. To this day, people have to cross contaminated streams and cannot use the groundwater for agriculture, bathing, or household needs. Given that red mud has to be stored for decades due to outdated techniques in Fria, these problems will persist even if the refinery will remain closed, which has been the case since 2012.

Despite considerable potential, as yet no African country has ever processed aluminum out of its own bauxite, and intra-African links are currently almost inexistent. Guinea is the only significant global bauxite producer that does not process this mineral (see Table 1.1). In 2008, the production price of bauxite in Guinea ranged from 10 to 15 USD in Débélé (CBK) to 20–25 USD per ton for the facilities of CBG in Kamsar and Sangaredi. With aluminum prices of 1500–2000 USD, this leads to a value added in the aluminum chain of less than 1% (17.5/1750), conceived as the value that is added to the input by the machinery and the workforce of a given production node.[8] This has contributed to a wide range of negative consequences, many of which will be treated in this book and could be summarized under the so-called "paradox of the plenty" (see Magrin 2013: 103; Ross 1999). Similar to almost 30 other countries in Sub-Saharan Africa (see Agnew and Grant 1996: 738), all the three significant African bauxite producer countries—Guinea, Ghana, and Sierra Leone—depend on only two raw materials for more than

Table 1.1 Bauxite production in Africa 1950–2010[a]

	1950	*1955*	*1960*	*1965*	*1970*	*1975*	*1980*	*1985*	*1990*	*1995*	*2000*	*2005*	*2010*
Guinea (% of worldwide production)	0	3	5	5	4	11	13	14	14	14	15	13	9
Guinea (% of African production)	7	80	86	78	76	89	92	90	90	96	97	96	91
Ghana (% of worldwide production)	2	1	1	1	1	0	0	0	0	1	0	0	0
Ghana (% of African production)	91	19	14	13	10	3	2	1	2	4	3	3	3
Sierra Leone (% of African production)	0	0	0	9	13	8	6	9	8	0	0	0	6
Zimbabwe, Mozambique, Tanzania (% of African production)	2	0	0	0	0	0	0	0	0	0	0	0	1
World production (Mt)	6110	17,521	27,033	36,847	56,873	73,608	89,216	84,168	112,908	112,229	104,000	133,478	191,428

[a]Cf. Knierzinger (2016): 208

75% of their export earnings. As the labor input and the spread effects of these sectors are considerably low,[9] this obstructs the development of labor-intensive industries and thereby also hampers democratization as governments continue to depend on the taxes of few transnational companies (TNCs), generated by several thousand miners (cf. Martin 2015). In addition, these taxes tend to remain low, because the dominating aluminum companies can shift their profits along their vertically integrated chains of production, ranging from bauxite production to semi-processed goods, such as aluminum cans (Diallo et al. 2011: 49). As most of government revenue in Guinea goes directly into salaries and/or foreign bank accounts, infrastructure development and other development programs are either dictated by international donors or undertaken more or less independently by the mining companies. At present, mining companies produce almost half of Guinea's electricity (Samb 2006: 19). Mainly due to risk reduction strategies of the mining companies and owing to geopolitical reasons, the low value added of mere bauxite production has been continuously decried but never overcome since the 1950s.

1.2 Bauxite and Aluminum Production in Africa

Africa is mainly involved in the global production network of aluminum by providing bauxite and raw aluminum for industries around the world. Out of Africa's five producer countries, only Guinea has significantly contributed to worldwide production since the 1950s. Since then, it has produced more than 80% of African bauxite. Other important mines existed and continue to exist in Sierra Leone, which exports the bauxite directly, and in Ghana, which produces both bauxite and aluminum, but has never managed to connect these production lines.

The only African alumina refinery is located in Fria, Guinea, and has produced about 1% of world output since the start of production in 1960. Apart from bauxite production in Guinea, Africa plays a significant part in the global aluminum industry mainly by providing energy and labor for smelters in Mozambique, South Africa, Egypt, Cameroon, Nigeria, and Ghana (see Tables 1.1 and 1.2 and Fig. 1.5). Because of their high demand for energy, these smelters have a considerable impact on the whole population of the countries, where they operate. They often attract huge investments in power plants, such as the Akosombo dam in Ghana or coal- fired power plants in Southern Africa, and they additionally play an important role in maintaining the base load of

Table 1.2 Aluminum production in Africa 1950–2010 (in Mt)[a]

	1950	*1955*	*1960*	*1965*	*1970*	*1975*	*1980*	*1985*	*1990*	*1995*	*2000*	*2005*	*2010*
Cameroon	0	0	48	56	58	57	43	90	93	79	86	90	76
Ghana	0	0	0	0	125	158	188	49	174	135	137	13	0
South Africa	0	0	0	0	0	84	87	165	159	195	673	846	807
Egypt	0	0	0	0	0	2	120	209	179	180	193	244	266
Mozambique	0	0	0	0	0	0	0	0	0	0	538	555	557
Nigeria	0	0	0	0	0	0	0	0	0	0	0	0	21
World Total	1640	3460	4950	6951	10,641	13,387	15,383	15,367	19,299	19,700	24,300	31,900	41,200

[a]This table has been published before in Knierzinger (2016): 208. Production in Nigeria started in the mid-1990s and stopped again in 2000 (Balzli and Herbermann 2000)

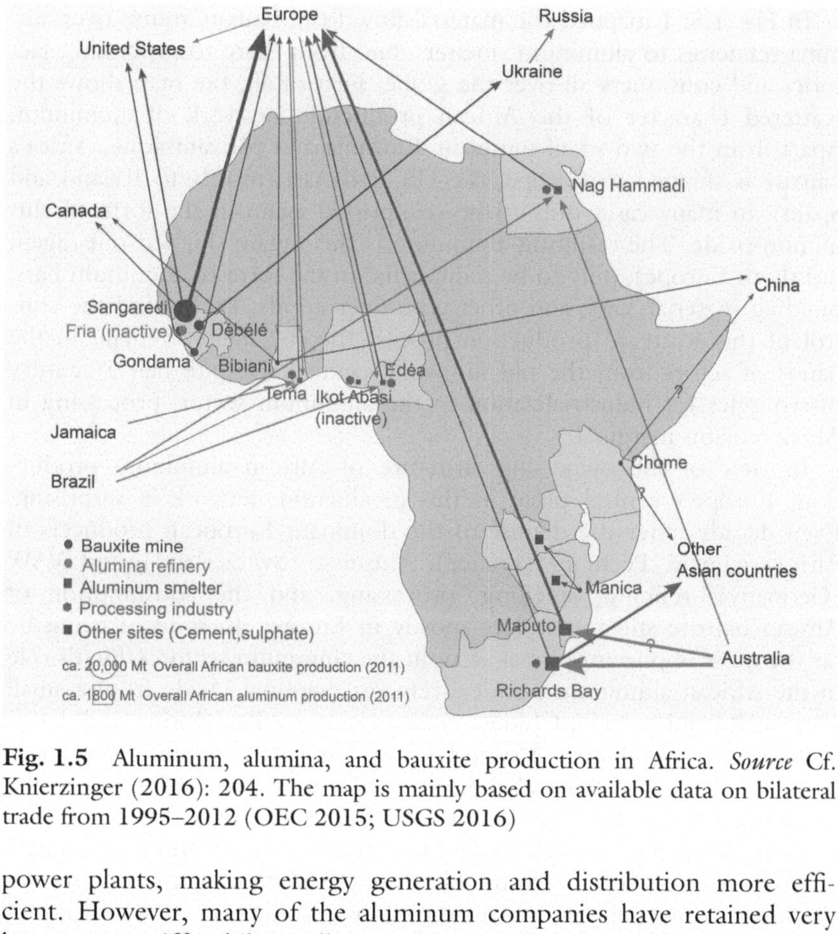

Fig. 1.5 Aluminum, alumina, and bauxite production in Africa. *Source* Cf. Knierzinger (2016): 204. The map is mainly based on available data on bilateral trade from 1995–2012 (OEC 2015; USGS 2016)

power plants, making energy generation and distribution more efficient. However, many of the aluminum companies have retained very low power tariffs while smaller companies and the population have been faced with frequent power cuts and escalating tariffs. In combination with the negative ecological impact of both smelters (fluor emissions) and power plants, this has led to protests in most of the affected countries, particularly in South Africa, Mozambique, Ghana, and Nigeria (cf. Custers 2013: 83–105). Mozambique is an example of a smelter with particularly low spin-off effects: the high demand for energy and the low demand for labor of the Mozal smelter have not yet been compensated for by, for example, the establishment of a processing industry based on the produced raw aluminum.

In Fig. 1.5, I mapped the material flow from bauxite mines over alumina refineries to aluminum smelters, and from there to processing factories and consumers all over the globe. First of all, the map shows the scattered character of the African production network of aluminum. Apart from the two small mines in Tanzania and Mozambique, Africa's bauxite is shipped to Europe, the US, and Asia (mostly to Ireland and Spain), in many cases only to be reimported again in the form of aluminum oxide. The resulting aluminum is then again shipped out (again mostly to Europe), only to be reimported in the form of aluminum bars, building material, cars, and other consumer goods. Profits and the control of this scattered production process thereby mainly remain in the hands of actors from the old industrial centers. Despite half a century of struggles for industrialization in the aluminum sector, processing in Africa remains marginal.

In view of the ownership structure of African aluminum production, Europe's central place in this production network is surprising. Even decades after the demise of the dominant European producers in Africa—namely, Pechiney (France), Alusuisse (Switzerland), and VAW (Germany)—refining, smelting, processing, and the consumption of African bauxite still takes place mostly in Europe. In spite of being by far the most important global actor in the aluminum sector, China's role in the African aluminum business remains marginal. Aside from a small mine in Bibiani, Ghana, Chinese enterprises are still exploring and developing sites or purchasing African bauxite. Similar to the alumina refinery of Fria, Bibiani had been abandoned by Western companies before the Chinese company Bosai Minerals Group stepped in (see below).

Apart from my research in Guinea, detailed information on bauxite mines in Africa is scarce. In the following, I will describe the situation in all African countries involved in the global production network of aluminum. In Chap. 3, the situation in Guinea will be analyzed in detail.

1.2.1 Tanzania, Mozambique, and Sierra Leone

The Tanzanian ore is mined in Chome, in the Same District which also includes the Shengena Forest Reserve, and then exported via Kenya. Production started in 2005 and the bauxite apparently goes to Asia. Local NGOs claim that the bauxite company, Willy Enterprise Ltd., is responsible for "escalating environmental degradation" and deforestation that threaten food production in the area and lead to water shortages

(Pesatimes 2013; Rainforest Rescue 2014). According to the District Natural Resources Officer of the Same District, the site employs 60 people from villages around the mine as security guards, drivers, cooks, and plant operators. The District Council's Executive Director underlined that revenues from bauxite mining can be used for road maintenance, health, and education services (ibid). According to USGS figures, Tanzania exported 130 Mt of bauxite in 2011. The Guinean CBG alone accounted in the same year for about 13,000 Mt (USGS 2016). The legal status of the bauxite mine in Chome is unclear. While several officials from the Ministry of Minerals seem to have approved the activities (Pesatimes 2013), the NGO Rainforest Rescue claims that the company violated Tanzanian law by not seeking approval by the land owners, and doubts the validity of its contract. According to the same NGO, the company also ignored a mining ban imposed by the government in June 2012 (Rainforest Rescue 2014).

The bauxite mine in Mozambique, Mina Alumina in Manica, is even smaller than the one in Chome and provides its ore to a cement company in Zambia and to a firm in Zimbabwe which produces aluminum sulfate for water purification (Mina Alumina 2015). Mozambique has more-or-less continuously exported several megatons of bauxite since the 1940s (USGS 2016). The mine is not connected to the Mozal aluminum smelter near the capital of Maputo that opened in 2000. Mozal belongs to BHP Billiton (47%), the South African parastatal Industrial Development Corporation (24%), Mitsubishi of Japan (25%), and the Government of Mozambique (4%). Export and import data provided by the OEC (2015) reveal that almost all of its raw aluminum is exported to various European countries, whereas the country reimports considerable quantities of processed aluminum. In 2010 alone, the export value of Mozambique's smelter reached almost 3 billion USD (ibid). The necessary aluminum oxide seems to come from Australia. In 2003, Mozal underwent a major expansion. Currently, the smelter requires 900 MW of electricity per year. Hathaway (2007) pointed out that the total consumption of the rest of Mozambique is just 300 MW. Only 7% of Mozambicans have access to electricity (ibid). The required power is delivered by state-owned companies in Mozambique, South Africa and Swaziland. Some of Mozal's electricity may originate from the Cahora Bassa dam (2075 MW) in Mozambique which is owned by the South African energy company Eskom. This company re-exports this electricity to Mozambique (ibid).

Sierra Leone exports its ore primarily to Romania, where it is processed into aluminum (Vimetco 2015; OEC 2015). Until 1995, production was controlled by the Swiss company Alusuisse. In 1961, it established Sieromco which acquired the rights over all bauxite reserves in Sierra Leone. In 1960, the Swiss company also bought 10% of the Guinean Fria, and in 1967, it acquired a smelter in Richards Bay, South Africa. In addition, it possessed a factory in Nigeria that produced tinware and tin roofs in the 1960s. Alusuisse's long-term director Emmanuel Meyer maintained excellent relations with Guinea's first president, Sékou Touré (Bauer and Maissen 1989: 53–63). Bauer and Maissen (1989: 109–139) present Sierra Leone in the 1980s as a bankrupt state that was highly dependent on Alusuisse. The industry had almost no linkages to the rest of the country and local elites addressed the company directly for infrastructural projects, thereby circumventing formal political institutions. About 95% of the 800 workers were organized in the United Mineworkers Union and received about 2 USD per day,[10] allowances and medical care included (ibid: 121). Alusuisse was bought by the Canadian company Alcan in 2000, which was in turn bought by Rio Tinto (Australia/GB) in 2007. Due to civil war from 1991 to 2002, the operations were shut down in 1995 and only taken up again in 2006 by Titanium Resources, which sold its assets to Vimetco in 2008. Vimetco is a Dutch enterprise with its management seat in Switzerland and further production sites in Romania and China (Vimetco 2008).

1.2.2 Mining and Smelting in Ghana

Ghanaian bauxite was discovered in 1914, although the first mine only became operative in 1941. Since then, Ghana has continually produced small amounts of bauxite. Similar to the French government at that time, the British had plans ("White paper" in 1955; see Riverton 2014: 32f) to build an integrated aluminum industry in order to escape the US dominance in the sector, but negotiations between British Aluminum (acquired by Reynolds in 1958), the Canadian, Alcan, and Kwame Nkrumah as prime minister of the Gold Coast failed. Similar to Guinea, Nkrumah turned to US companies (Kaiser and Reynolds) after independence in 1957 as an escape from colonial domination (see below). In 1973, the Ghanaian government took over 55% of the bauxite mine of Reynolds and formed the Ghana Bauxite Company in Bibiani, Awaso

(Asamoah 2014: 224). In 1982, the remaining stakes of Reynolds were sold to Alcan, which was acquired by Rio Tinto in 2007. Three years later, Rio Tinto sold its 80% shares of the facilities in Ghana to Bosai Minerals Group of China, apparently mostly due to the decline of Ghana's Western railway which had facilitated the transport to the coast in Takoradi, about 240 km from the mine (Rio Tinto Alcan 2010). In 2012, the new Chinese owners even resorted to transporting the ore to the coast with trucks (Oxford Business Group 2014: 144).

The takeover of Bosai was criticized by the Ghanaian media. Announcements of payments and the received sums did not match and considerable amounts of bauxite were immediately exported to China despite Ghanaian legal provisions stipulating that this change of exports had to be authorized by the parliament (Dogbey 2010). In a memorandum of understanding in 2010, Bosai also announced the construction of an alumina refinery and even of a Ghanaian smelter. The refinery should have been finished in 2014 (Biswas 2012). According to Rio Tinto, the bauxite mine in Bibiani in Western Ghana employed 299 people in 2008. Interestingly enough, the official investment in local development (health, education, water, and sanitation) amounted to 100,000 USD per year until 2012, the same sum that is annually paid out by the Compagnie des Bauxites de Guinée (CBG) in Guinea, where Rio Tinto still holds one of the three dominant shares (Bertolli 2010).

Ghana's bauxite mines have never been connected to the Ghanaian aluminum smelter in Tema. After independence in 1957, Ghana's first president Kwame Nkrumah directly negotiated with Henry Kaiser, the head of the US company Kaiser Aluminum. Kaiser initially agreed to produce aluminum out of Ghanaian bauxite, but lowered the expectations of the Ghanaian president several times during the negotiations. The initial plan of Nkrumah covered a fully integrated aluminum industry (from bauxite to aluminum processing including chemical facilities), countrywide electrification, irrigation schemes, and a local fishing industry. Kaiser disagreed on the last two improvements and postponed the idea of an integrated aluminum complex. In spite of significant bauxite reserves only 50 km away from the dam in Kibi, the company wanted to import bauxite at least for the next 10 years. According to Wiederstein (1994: 35–38), Kaiser also presented completely unrealistic financial plans. In the end, it became clear that Kaiser was only able to provide funds for the construction of the smelter (10% of the capital came from Reynolds), with a loan from the US export bank, while the dam, which was almost twice as expensive, had to be financed by the Ghanaian government, the World Bank and the US government. Ghana additionally had to sign an agreement that it would not expropriate the company. Soumah maintains that the World Bank president from 1963 to 1968, George D. Woods, had been working as a financial advisor for Kaiser before taking office in Washington (Soumah 2008: 146). The price of electricity for Kaiser and Reynolds was particularly low. In the early 1980s, when energy prices had already increased ninefold since the start of production due to the oil crises, the power rate for the aluminum companies had only been doubled (Riverton 2014: 45). The relations between the government of Ghana and the US worsened considerably because of these developments and Nkrumah, again similar to Touré, tried to involve the Soviet Union in constructing a second dam in the area that is now the Bui National Park on the borders to Ivory Coast.[11] Kaiser was not pleased and is said to have used his contacts to the head of the CIA John McCone and the Rockefellers (and probably also George D. Woods from the World Bank) to bring down the uncomfortable president. Nkrumah's political fall shortly after the inauguration of the Akosombo dam was, therefore, hardly a coincidence, more so if one considers that, according to a New York Times article shortly after the coup, the CIA "advised and supported" the rebels (Riverton 2014: 34). The new government was cooperative and continued to provide subsidized energy, which made it even more difficult to repay the debts that had been created by the construction of the dam. Even today, Ghanaian bauxite mining and Ghanaian electrolysis in Tema are not connected to each other. In the last 20 years, the Ghanaian bauxite has primarily gone to Germany, Greece, the UK, Canada, and China, while the smelter in Tema imports alumina from Latin America (mostly Jamaica) and again from Europe (see Fig. 1.3).[12]

The dam itself has been described as an ecological catastrophe by various researchers. 78.000 people were forced to relocate and 60% of them left their new homes again after a few years. In the same period, rural economies in the region collapsed, leading to more migration and an increased crime rate. 8400 km^2 of rain forest were flooded by the dam's construction, creating a water body more than eight times larger in surface than the Three Gorges Dam. This contributed to a disruption of the rainy seasons in the region, and led to the spreading of diseases, such as schistosomiasis and malaria, as well as to the erosion of the coast at the Volta estuary (Riverton 2014: 41). Due to recurrent draught, the water level of the dam was repeatedly too low to produce a sufficient energy for aluminum smelting.[13] Between 1982 and 2004, the plant shuts down 13 times because of power shortages. In 2004, Kaiser sold its 90% stakes to the government, and in 2008, the plant was fully nationalized (Alcoa 2015). From 2007 to 2011 the plant was continually shut down. In 2013, Ghanaian authorities launched the "Integrated Aluminium Project" that is intended to link Ghanaian bauxite with Valco. At the time of writing, however, no investors have been found (Adjorlolo 2015).

1.2.3 Mining and Refining in Guinea

Guinea became part of the global production network of aluminum in two steps. The first was already prepared before independence and led to the construction of the company town of Fria, which was built under French leadership and remained under French control (who worked with a variety of Western shareholders) until 1997 when both the French company Pechiney and its factory in Fria encountered problems. After 3 years of state ownership, US company Reynolds assumed control, before it was sold to Rusal in 2006. At present, Fria is the only alumina refinery on African soil, though because of a lockout by Rusal in April 2012 (see Chap. 3.1), it is no longer operational. Its alumina was initially mostly shipped to Cameroon (Edéa), where it was transformed into aluminum, which was then mainly sold to French and West African producers. Beside Fria, Rusal currently controls another bauxite mine in Débélé, near the city of Kindia. This fully state-owned facility, the *Compagnie des Bauxites de Kindia* (CBK), began production in 1974 with the assistance of the Soviet Union. It is located near the city of Kindia and exports almost entirely to Russia.

Besides this Russian controlled company network, Guinea also hosts a Western joint venture, called *Compagnie des Bauxites de Guinée*, which began production in 1973. The bauxite of this network is mined in Sangaredi and processed in the coastal city of Kamsar. This mine is by far the greatest economic asset of Guinea. It is managed by Alcoa in a joint venture with Rio Tinto (both holding 22.95%), the Guinean state (49%) and Dadco, a company registered on Guernsey, which produces alumina in Germany (see Fig. 1.6).

These two company networks—the Russian assets (Fria and Débélé) and the US-American and Australian (CBG in Sangaredi and Kamsar)—form two corporate decision spaces that overlap and conflict with the formal political system in Guinea. Through their taxes and the taxes of

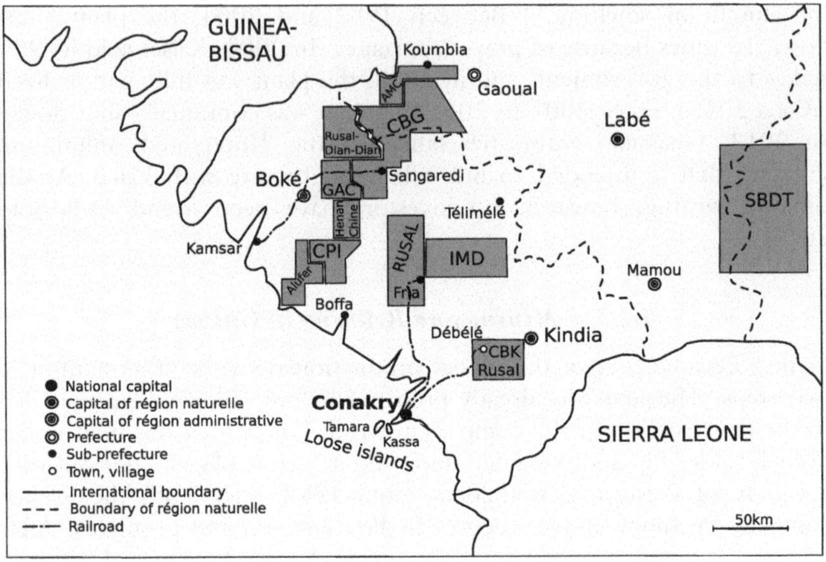

Fig. 1.6 Bauxite mining permits in Guinea. *Source* Ministère des Mines et de la Géologie (2013)

their workers, they are responsible for a large part of Guinean public revenue, and thereby provide for public salaries, social services, and military expenditure. The incomes of the workers and the subcontractors additionally end up in the pockets of many other Guineans, providing agricultural products, craft, and services in the mining towns. In addition, these companies almost exclusively run the cities, where they produce: they provide water and electricity, maintain roads, administrational, religious and apartment buildings, finance cultural, and recreational facilities, and provide social services, such as medical care. In Chap. 3, I will return to these mining towns and their connections with other African production sites.

1.2.4 *Alucam in Cameroon*

Right from the start of production, Fria delivered a good part of its alumina to Alucam in Edéa, which was the first aluminum plant on Africa soil. According to Alucam's first president (and at the same time vice managing director of Pechiney),[14] Jean Matter, the smelter in Edéa was built on rather short notice as part of the French efforts to substitute the shrinking imports from the US in the early 1950s (Matter 1959: 3; see also Pauthier 2002: 27; Pezet et al. 2009: 19). The project was initiated in 1953,[15] and started production in 1957. It reached full capacity (about 50,000 tons per year) in 1958, before the completion of the alumina refinery in Fria (Vitry 1957), and was thereby the first real investment of Pechiney outside metropolitan France (Pezet et al. 2009: 28). The first Cameroonian regime was economically liberal and Francophile. In his independence speech, the first president, Ahmadou Ahidjo, addressed the former colonial power more often than the Cameroonians (Ahidjo 1960: 2).

After independence, the Alucam belonged to Pechiney (48%), the French state (*Caisse Centrale de Coopération Economique*, 10.7%) and the Cameroonian state (35%; Campbell 1986: 54). From 1958 to 2012, for more than 50 years, Alucam thereby received most of its alumina from Fria in Guinea. OEC data show that from 1995 to the early 2000s, the alumina came almost exclusively from Guinea and was then replaced by Brazilian production (OEC 2015). The primary aluminum produced in Edéa was then not exclusively exported again, but further processed by Socatral and then used in various factories to produce finished goods. Socatral was created in 1963 as an affiliate of Alucam and remains the largest producer of rolled aluminum in West and Central Africa (23,000 tons per year in 2007, of which only 30% is exported; Pechiney Ugine Kuhlmann 1979). In 1979, it counted 750 employees and received two-thirds of Alucam's output (Bourgois 1979). The rest of the primary aluminum went to Europe and by the 1970s also to China (Debost 1977: 4; Pechiney Ugine Kuhlmann 1977: 6). Most of the African processing factories were subsidiaries of Alucam. Among them was Alubassa, which is still producing for Alucam today.[16] The finished products included roofing, building materials, and household gadgets, which were then sold mostly in Cameroon, the Central African Republic, Ivory Coast, Senegal, and the two Congos (Pechiney Ugine Kuhlmann 1972: 2; Senghaas-Knobloch 1975: 286; see also Habig 1983: 154).[17]

Within the boundaries of West Africa—the preferred territorial entity of the early leaders of African independence (see Rühl 1959)—integrated production—therefore, became a reality for more than 50 years. Until recently, a significant part of the Guinean bauxite was processed into roofing, building materials, and household gadgets and even went into the recycling loop of West Africa (e.g., for artisanal cooking pot production in West Africa, see Osborn 2009). In 2002, Socratal received 25% of the Cameroonian raw aluminum production (Girrbach and Andrianarisata 2003). Despite these comparably sound industrialization plans and practices, the project still faced opposition among local small firms and anti-colonialist movements. According to Maurice Laparra, director of the factory in Edéa since the 1960s, this site was mainly chosen, because the necessary power plant already existed since 1947 (*Electricité de France* (*EDF*) on the river Sanaga) and had abundant excess capacity (Interview with Former manager 15/09/2011; Pauthier 2002: 27). Its power was nonetheless quintupled in 1958 using Alucam as its base load customer (Husband et al. 2009: 30). When production started, the smelter consumed 50 times more energy than the largest city in Cameroon, Douala, with a population of about 250,000 at that time (Matter 1959: 3). Like the smelters in Ghana and Mozambique, the Alucam aluminum smelter remains the single largest consumer of energy in the country (Hathaway 2007). In 2012, it consumed between 40 and 50% of the electricity that was produced in Cameroon (Belengue 2014). According to Hathaway, Alucam profited from highly subsidized rates for electricity consumption, which have continued to today. Alucam buys 40% of the power produced by AES-Sonel, compared to 25% purchased by residential and other users, yet Alucam's payments make up only 12% of AES-Sonel's revenue compared to 54% from other consumers. Only 20% of the Cameroonian population have access to electricity (Hathaway 2007).

Pauthier writes that Fria was a "social and humanitarian success story" in comparison with Edéa, which was also built by Pechiney and which she characterizes as a severely segregated and heavily guarded state in the state. Initially, only 335 employees were planned: 135 Europeans and 200 Africans, all of them men (Pauthier 2002: 14, 27). The housing estates comprised three cites—one being reserved for the Europeans—a sports court, a kindergarten, a *centre de formation ménagère* (a housekeeping school), a *centre médico-social*, and a *centre commercial.* The racist segregation was apparently striking: not only were the districts separated, but also the schools, the clubs of every *cité* and the sports teams. The only swimming pool was to be found in the European *cité* (Pauthier 2002: 27; Pechiney Ugine Kuhlmann 1982: 4–5).

The high percentage of expatriate employees was an issue from the start. The first Cameroonian foreman was hired in 1962 and in 1970 Alucam got its first Cameroonian executive (Pezet et al. 2009: 31).

In 2005, Alcan put pressure on the Cameroonian government to expand its energy production by constructing two new dams, Lom Pangar and Nachtigal. Support for the Lom Pangar dam was expected from the AFD and the World Bank. Nachtigal was to be financed, developed, and operated by Alucam along with a triplication of the existing aluminum smelter capacity, development of associated infrastructure, and possibly bauxite mining. Alcan made it clear that without the dams, it would eventually leave the country (Hathaway 2007). When Alcan was taken over by Rio Tinto in 2007, the new owners continued and expanded these projects, heralding investments of over 5 billion USD (Belengue 2014), only to suddenly announce the selling of its 46.67 stake in Alucam in October 2014. Another 46.67% is currently held by the Cameroonian state, 5.6% by the *Agence française du développement,* and 1.1% by the Alucam workers (ibid). The Cameroonian *Réseau Associatif des Consommateurs de l'Energie* explained this change of mind as resulting from disagreements between the Cameroonian government and Rio Tinto over the development of the Nachtigal dam. Three months before the turnaround of Rio Tinto, on 10 July 2014, the Cameroonian government awarded the construction of Nachtigal to the French EDF (*Electricité de France*) instead of Alcan's successor (ibid).

Similar to Guinea, Cameroon became a preferred target of the global aluminum industry during the last raw material boom. Among these projects is Cameroon Alumina Limited, a joint venture of Hindalco (45%, India), Dubal (45%, Dubai), and Hydromines (10%, USA) that aim at mining and processing bauxite into aluminum oxide. In 2014, Dubal merged with Mubadala (Abu Dhabi) into Emirates Global Aluminium, which thereby became one of the five largest aluminum companies in the world. Emirates Global Aluminium also acquired the Guinea Alumina Corporation (GAC), a large aluminum project in Guinea.

1.2.5 Nigeria's "Alu City" Ikot Abasi

The Nigerian Alscon was built by the German company Ferrostaal (20%), by Reynolds (10%), and the Nigerian state (70%) in the mid-1990s, during the dictatorship of Sani Abacha, in the Niger Delta with the aim to use formerly flared gas from oil drillings (Balzli and Herbermann 2000; Abasiattai et al. 2013b). However, the plant was only in operation for a few months, from the end of 1998 to May 1999, when it ceased operations, mainly due to inadequate gas supply. In the same year, the new government Obasanjo put the plant up for sale under its privatization program (Udo 2015). Several reports (e.g., Walker 2015) stating that Alscon processed bauxite from nearby mines do not seem to be true (cf. USGS 2016; Abasiattai et al. 2013a). In 2007, Rusal acquired a majority stake, relaunched the factory in 2008, and closed it again in 2013, due to overcapacities of the company and following a decision of the Nigerian Supreme Court in July 2012 to strip Rusal of its stakes. The smelter processed alumina from Fria (Interview with Former executive Rusal 25/03/2014). The Nigerian Supreme court based its decision on corruption claims concerning the acquisition of the plant by Rusal (Udo 2015). Two years after the lockout in the Guinean Fria, Rusal was thereby at the center of yet another shutdown of a whole city. Nigeria's "Alu City" Ikot Abasi had over 1800 direct employees and provided work for another 2500 from support-service companies. In September 2014, the electricity for this agglomeration with more than 100,000 inhabitants, which had been furnished as a token of social responsibility by Alscon, was cut (ibid). Similar to the situation in Fria, the workers of Alu City accused the management of Rusal of "economic sabotage" in Ikot Abasi (Ekpimah 2014). As the takeover corresponded with the global financial crisis and Rusal's ensuing debt crisis, the factory never exceeded 12% of its 193 metric tons capacity (Udo 2015). Detailed information concerning the social consequences of this shutdown is not available.

1.2.6 South Africa and Egypt

Besides these seven countries directly linked to African bauxite production, the two remaining smelters in South Africa and Egypt import alumina from South America and Australia. The first smelter in South Africa—the Bayside plant—was constructed in 1969 by Alusuisse and started production in 1972. By then, Alusuisse had a 22% share in the smelter, while the majority of the shares were held by the South African "Industrial Development Corporation" (van der Walt 2009). Today, two smelters in Richards Bay—Bayside and Hillside—belong to BHP Billiton (BHP Billiton 2009). Their ownership structure as well as their geographical proximity make Mozal and Richards Bay one transnational aluminum cluster with an output of about 1400 Mt of raw aluminum per year (cf. USGS 2016). Bayside and Hillside were built under South Africa's apartheid regime and have power agreements that are adjusted to the international market price of aluminum. If aluminum prices drop significantly, electricity prices are below the cost rates (Hathaway 2007). The needed aluminum oxide comes from Australia (OEC 2015).

The recent raw material boom also provoked the announcement of numerous projects in the South African aluminum sector. In 2007, Rusal considered the opening of an aluminum smelter with an expected energy consumption of 1300 MW (Hathaway 2007). In 2006, Alcan (which was bought by Rio Tinto 1 year later) announced the construction of a smelter in Coega with a power consumption of another 1300 MW and investments of 2.7 billion USD. The smelter would have been the cornerstone of a South African development zone by the same name. After the South African government had spent up to one billion USD on developing the industrial zone (cf. ibid), Rio Tinto Alcan scrapped its plan to build the smelter due to the start of a countrywide electricity crisis in 2008 (Mail and Guardian 2009) which has not been resolved at the time of writing. While Bayside was shut down in July 2014 due to power shortages and is about to be dismantled, the continuing subsidized energy consumption of Hillside and Mozal (which also receives electricity from the South African Escom) remains under public criticism (Custers 2013: 83–105; Moorcroft 2015).

Egypt's only smelter is in Nag Hammadi, 100 km North of Luxor. It started production in 1975 and has steadily increased production since reaching 260 Mt per year in 2010 (USGS 2016). The facilities are equipped with a rolling mill and a casting house, enabling the processing of a good part of raw aluminum production. As in many other African production sites, the construction of the smelter was accompanied by the erection of a whole city with a hospital, a stadium, several sport courts, swimming pools, an open air cinema, a theatre, etc. The company itself describes the city as "being virtually self-sufficient". It has its own farms and plantations and obtains water via more than 30 "artesian wells" (Egyptalum 2015). Egyptalum is state-owned, receives its energy from the Aswan dam, and exports mainly to Europe (cf. OEC 2015).

1.3 Concluding Remarks

This short outline of the African production network first of all shows a great variety of cases that are weakly linked to each other. Some mines and smelters have existed since the 1950s, while others have just been created and virtually all of the facilities—probably with the exception of those in Egypt and South Africa—are enclaves that are more strongly linked to factories and offices outside the African continent than to their regional surroundings. From this perspective, impressively little has changed since the implantation of the industry on the continent in the 1950s and 1960s. At the eve of African countries gaining independence, both the population and the political leaders strongly believed that everyone would have access to running water, electricity, and roads within a few years. In almost all of the aforementioned countries, these blessings were to be brought about at least partly by the erection of bauxite mines and smelters. Besides the integrated production of aluminum out of bauxite—which would have been possible in many of these countries—political strategists had equally great expectations concerning the processing of aluminum and the ancillary use of the large dams. The newly-created water bodies were to be used for irrigation schemes and whole fishery industries and the newly-installed power stations were to provide energy to electrify homes across the African continent. Most of these plans have proven untenable. Lacking control over critical technical know-how, the new leaders had the choice to cooperate directly with either international corporations or capitalist and communist governments, or both. In all of these cases, the interests of these new or old partners conflicted with the plans of the new nationalist leaders. The European aluminum industry had been in decline since the 1970s, struggling with high energy prices and rising environmentalism. The concerned European governments intended losing neither value added produced by aluminum smelters nor jobs in the industry by helping to advance the transformation of bauxite in Africa. On the other hand, the aluminum companies were afraid of expropriation and thus constructed a complicated network of production facilities that should minimize the negative consequences of possible nationalizations. In the long run, this led to a continuation of the international division of labor, with Africans remaining providers of raw materials. Today, these industrialist dreams seem even less realistic than during the so-called "lost decade" of the 1980s. Changing power relations within the global production network of aluminum have further reduced the few intra-continental linkages that previously existed, such as the one between Guinea and Cameroon. The rise of new competitors in the aluminum market from Asia, South America, and Oceania has led to an even stronger spreading of the network with new stopovers of the material flow on other continents. At the same time, large dams for energy production—an essential part of plans for integrated production of aluminum on the African continent until now—have come under hefty criticism in recent decades, not only due to their immediate negative effects, such as vast relocation measures, changes in regional climates, the proliferation of diseases, etc., but also owing to technical reasons: Strong changes in precipitation lead to marked differences in the dams' water levels, which in turn have proven to affect the efficiency of the turbines much more than originally projected. Moreover, recent studies show that dams might be a huge source of methane emissions and thus strongly accelerate global warming.

The recent raw material boom neither significantly augmented the profitability of aluminum production—during the boom many observers thought that the world might enter a new phase of permanently high raw material prices—nor did it lead to a shift of power relations. New investors from China, the Arab peninsula, and South America have taken part in something that has repeatedly been called a "New Scramble for Africa", although most of these interventions have remained restricted to the speculative and strategic acquisition of claims rather than new mines and factories. In addition—and again contrary to the hopes of many commentators—this proliferation of actors has hardly led to a stronger bargaining position of African governments (and in turn African civil society). Instead, the "politics of the belly" (Bayart 1989) seem to have found various new providers and coherent industrial planning has become even less feasible. However, this combination of relatively high economic growth and little social progress has nonetheless seemed to trigger a new wave of indignation against extractivism all over Africa. Marked by the experiences of the mining bust resulting from the global financial crisis, a new generation of sons and daughters of miners is about to demand "a future without mines", as formulated by a social movement in Fria. In the following chapter, the history of this and three other Guinean bauxite mining towns will be presented in detail.

Notes

1. In his 420-page book on "Fria en Guinée", Larrue praises Fria as "an act of faith [and] hope", "a common ground for the people" and a "pledge for the future of Guinea" (Larrue 1997: 14; translation by JK). Nonetheless, Larrue's subtle critique of (neo-)colonialist pathos and his inquiries concerning the planned installation of a smelter in Fria in the

1950s (Larrue 1997: 90ff) made the lecture of this book one of the most important sources on Fria.

2. Similar shifts have also occurred in the worldwide production of copper, coal, lead, nickel, zinc and tin (Huy 2014: 9).
3. Russian production began in 1932 and reached about 10% of world production in the early-1940s (Rusal 2013).
4. The dominant Japanese companies Showa Denko and Nippon Light Metal were founded in 1934 and 1940 (Bunker and Ciccantell 1995: 58).
5. For aluminum as a strategic war material until the 1960s, see Bauer and Maissen (1989): 37–53; Bunker and Ciccantell (1995): 58; Farin and Reibsamen (1969): 13; Otte (1998): 25–29; Wiederstein (1994): 13.
6. Custers and Matthysen (2009): 47; Roberts et al. (2008).
7. Based upon this study of the Austrian ministry of health, the Austrian armed forces have recently banned the use of aluminum crockery.
8. Cf. Girvan (1971), Graham (1982): 78; Habig (1983): 85; Forster (1976): 133.
9. Artisanal gold mining provides many jobs but occupies rather low importance for the country's exports and governmental income.
10. Today, this would amount to about 4 USD per day or about 100 USD per month.
11. The construction of the Bui Dam finally started with Chinese assistance in 2009.
12. Riverton (2014): 36, 45, OEC (2015), cf. Dogbey (2010).
13. Wiederstein (1994): 35–38, Lanning and Mueller (1979): 429–435, Riverton (2014).
14. Pezet et al. (2009): 19.
15. Matter (1959): 3.
16. Carbonell (2007); cf. Husband et al. (2009): 30.
17. In 1971, Pechiney received 220 Mt of the 700 Mt of alumina being produced at Fria. About one-half of that, 104 Mt, was sent to the smelter in Edéa, which produced 54 Mt of aluminum. Of this, 34 Mt was exported and 15 Mt went to Socatral to be processed into 9.8 Mt of aluminum sheets (*tôles*), 1.4 Mt of *disques* and 3.8 Mt of *bandes*. The sheets were then sold in Cameroon (8500 t) or in the rest of Africa (1300 t). Of the *disques* and *bandes*, Alubassa received 700 t, Sozalu (Zaire) 50 t, C.B.I. (Central Africa) 160 t, IVORAL (Ivory Coast) 4150 t, Senegal 60 t and C.P.N. (Congo-Brassaville) 130 t. All these companies were subsidiaries of PUK and produced solely household articles (Pechiney Ugine Kuhlmann 1972: 2).

Bibliography

Abasiattai, Monday B., Violetta I. Ekpo, and Chukwurah Ezebube 2013a. *Important Raw Material: Alumina.* https://alscon.net/ikot-abasi-smelter/76-important-raw-material-alumina.html. Checked on Sept 22, 2015.

Abasiattai, Monday B., Violetta I. Ekpo, Chukwurah Ezebube. 2013b. *Philosophy of the Ikot Abasi Aluminium Smelter Plant.* https://alscon.net/ikot-abasi-smelter/70-phylosophy-of-the-ikot-abasi-aluminium-smelter-plant.html. Checked on Sept 22, 2015.

Adjorlolo, Ruth Abla. 2015. VALCO Committed to Adding Value to Ghana's Bauxites. In *Ghana Broadcasting Corporation*, Jan 22, 2015, http://m.gbcghana.com/1.1965403. Checked on Sept 24, 2015.

Agnew, John, and Richard Grant. 1996. Representing Africa: The Geography of Africa in World Trade, 1960–1992. *Annals of the Association of American Geographers* 86 (4): 729–744.

Ahidjo, M. Ahmadou. 1960. Cameroun Indépendant. *Bulletin Alucam* 10: 1–2.

Alcoa. 2015. *Corporate Overview.* http://www.alcoa.com/global/en/about_alcoa/overview.asp. Checked on Mar 21, 2015.

Asamoah, Obed Yao. 2014. *Political History of Ghana (1950–2013): The Experience of a Non-Conformist.* Bloomington: AuthorHouse.

Balzli, Beat, and Jan Dirk Herbermann. 2000. Berüchtigte Kundschaft. In spiegel.de, May 29, 2000. http://www.spiegel.de/spiegel/print/d-16525418.html. Checked on Oct 16, 2014.

Bauer, Tobias, and Toya Maissen. 1989. *Silbersonne am Horizont: ALUSUISSE - eine Schweizer Kolonialgeschichte.* Zürich: Limmat-Verlag.

Bayart, Jean-François. 1989. *The State in Africa. The Politics of the Belly.* London and New York: Longman.

Belengue, Jean Baudelaire. 2014. La raison du désengagement de Rio Tinto Alcan d'Alucam!, Oct 13, 2014, http://www.camnews24.net/fr/economie/cameroun/47327-la-raison-du-desengagement-de-rio-tinto-alcan-dalucam. Checked on Oct 16, 2014.

Bertolli, Stefano. 2010. Rio Tinto Alcan Sells Its Share in the Ghana Bauxite Company. http://www.riotintoalcan.com/ENG/media/media_releases_1796.asp. Checked on Sept 21, 2015.

Billiton, B.H.P. 2009. Aluminium Industry in Southern Africa—Aluminium Scuba Diving Cylinders. http://www.ali-tek.com/aluminium.htm. Updated on Aug 13, 2009, checked on May 27, 2013.

Biswas, Beethika. 2012. The Untapped Bauxite Potential in Africa: Ghana, Nigeria, Cameroon and Sierra Leone. In *AlCirlce*, Sept 25, 2012. http://blog.alcircle.com/2012/09/25/the-untapped-bauxite-potential-in-africa-ghana-nigeria-cameroon-and-sierra-leone/. Checked on Sept 21, 2015.

Brinkmann, Manfred. 2004. Silber aus Lehm. Wie nachhaltig ist Aluminium? Aachen: Brot für die Welt.

Bourgois, Catherine. 1979. Alucam se développe au Cameroun. *Jalons* 24: 16.

Bund der Energieverbraucher. 2015. Tipp 63: Wenig Aluminium verbrauchen. http://www.energieverbraucher.de/de/tipp63-aluminium__2047/.

Bundesministerium für Gesundheit. 2014. Aluminium—Toxikologie und gesundheitliche Aspekte körpernaher Anwendungen, Wien. http://www.bmg.gv.at/cms/home/attachments/3/9/6/CH1146/CMS1402477436830/aluminium_studie_2014.pdf. Checked on Mar 25, 2015.

Bunker, Stephen G., Ciccantell, Paul S. 1995. The Evolution of the World Aluminum Industry. In *States, Firms, and Raw Materials. The World Economy and Ecology of Aluminium*, ed. Bradford Barham, Stephen G. Bunker, and Denis O'Hearn. 39–62. Madison: University of Wisconsin Press.

Campbell, Bonnie, and Jennifer Clapp. 1995. Guinea's Economic Performance Under Structural Adjustment: Importance of Mining and Agriculture. *The Journal of Modern African Studies* 33 (3): 425–449.

Carbonell, Mauve. 2007. Aluminium in Cameroon: Markets and Uses. In *Alucam, An African Destiny: 50 Years of Aluminium in Cameroon, 1957–2007*, ed. Maurice Laparra, and Ivan Grinberg, 125–148. Paris: REF.2C editions.

Custers, Raf. 2013. *Chasseurs de matières premières*. Bruxelles: Couleur livres.

Custers, Raf, and Ken Matthysen 2009. Africa's Natural Resources in a Global Context. Antwerp.

Dadco. 2013. *Dadco Alumina & Chemicals Ltd*. http://www.dadcoalumina.com/. Checked on May 17, 2013.

Debost, Jean. 1977. La guenon, la locomotive et la bauxite. *Aluminium Pechiney* 31: 4–9.

Diallo, Mamadou Chérif, et al. 2011. *Les enjeux de la gouvernance du secteur minier en Guinée*. Conakry: Cooperation Guinée Allemagne.

Dogbey, Larry-Alans. 2010. *British Duped Ghana Millions of Dollars in Bauxite Sale to Chinese*. In *GhanaWeb*, July 13, 2010. http://www.ghanaweb.com/GhanaHomePage/NewsArchive/British-Duped-Ghana-Millions-Of-Dollars-In-Bauxite-Sale-To-Chinese-186000. Checked on Sept 21, 2015.

Egyptalum. 2015. *Home Page*. http://www.egyptalum.com.eg/. Checked on Sept 30, 2015.

Ekpimah, Etim. 2014. Labour Union Protests ALSCON's Shutdown. In *The Punch*, Apr 01, 2014. http://www.punchng.com/metro-plus/labour-union-protests-alscons-shutdown/. Checked on Sept 22, 2015.

Farin, Philip, and Gary G. Reibsamen. 1969. *Aluminum. Profile of an Industry*. New York: Metals Week.

Forster, Meinhard. 1976. Struktur und Risiken der deutschen Nichteisen-Metallversorgung dargestellt am Beispiel Kupfer, Aluminium, Nickel, Zink, Blei und Zinn. Hamburg: Weltarchiv.

Girrbach, Elisabeth, and John Andrianarisata. 2003. *Pechiney—Alucam Cameroon Business Case Study. Global Compact Learning Forum*. Eschborn, Antananarivo: GTZ.

Girvan, Norman. 1971. *Foreign Capital and Economic Underdevelopment in Jamaica*. Kingston: Institute of Social and Economic Research, University of the West Indies.

Graham, Ronald. 1982. *The Aluminium Industry and the Third World. Multinational Corporations and Underdevelopment*. London: Zed Press.

Graham-Rowe, Duncan. 2005. Hydroelectric Power's Dirty Secret Revealed. February 2005 by Duncan. In newscientist.com, Feb 24, 2005. http://www.newscientist.com/article/dn7046-hydroelectric-powers-dirty-secret-revealed.html. Checked on Mar 23, 2015.

Grinberg, Ivan. 1994. *Le patrimoine, nouvelle danseuse pour l'entreprise? Le cas de Pechiney*. Paris: IHA.

Habig, Gutmann. 1983. *Möglichkeiten und Grenzen einer Kontrolle internationaler Rohstoffmärkte durch Entwicklungsländer. Das Beispiel des Kupfer- und Aluminiummarktes*. Hamburg: Weltarchiv.

Hathaway, Terri. 2007. Aluminum in Africa. A Case Study for Earthlife Africa eThekwini and Friends of the Earth. http://www.internationalrivers.org/resources/aluminum-in-africa-3998. Checked on Sept 27, 2015.

Hildebrand, Lars. 2007. Die globale Güterkette der Aluminiumindustrie. Weltmarktintegration als Entwicklungsstrategie? Erfahrungen aus Brasilien. Diplomarbeit. Hamburg: Universität Hamburg.

Husband, Charles, Gary McMahon, and Peter van der Veen. 2009. *The Aluminum Industry in West and Central Africa. Lessons Learned and Prospects for the Future. Extractive Industries and Development Series* #13. Washington, D.C: World Bank.

Hütz-Adams, Friedel, Ann-Kathrin Voge, Sascha Liese, Stefan Hörmann, and Jan Per Bethge. 2014. Nachhaltige Rohstoffe für den deutschen Automobilsektor. Herausforderungen und Lösungswege. Bonn: Südwind e.V.

Huy, Dieter. 2014. *Deutschland —Rohstoffsituation 2013*. Hannover: Bundesanstalt für Geowissenschaften und Rohstoffe.

Institut pour l'Histoire de l'Aluminium. 2015. Officers and Board of Directors. http://www.histalu.org/english/iha-rubrique-article.php?rub=2&art=155. Checked on Feb 22, 2015.

Kerkow, Uwe, Jens Martens, and Axel Müller. 2012. Vom Erz zum Auto. Abbaubedinugungen und Lieferketten im Rohstoffsektor und die Verantwortung der deutschen Automobilindustrie. Aachen, Bonn, Stuttgart: MISEROR, "Brot für die Welt", Global Policy Forum Europe.

Knierzinger, Johannes. 2016. Bauxit und Aluminium aus Afrika: Ausbeutung auf Umwegen. In Globale Ressourcen und Rohstoffpolitik. Aktuelle Auseinandersetzungen im historischen Kontext. Historische Sozialkunde/ Internationale Entwicklung 35, ed. Karin Fischer, Johannes Jäger, and Lukas Schmidt, 203–220. Wien: New Academic Press.

Lanning, Greg, and Marti Mueller. 1979. *Africa Undermined. Mining Companies and the Underdevelopment of Africa*. Harmondsworth: Penguin (A pelican original).

Laparra, Maurice. 1995. L'excellence technologique, outil de développement international: l'aluminium de Pechiney. *Défense* 69: 43–46.

Larrue, Jacques. 1991. Fria en Guinée. Des aspects humains d'une industrialisation differente. *Cahiers d'histoire de l'aluminium* 7: 37–49.

Larrue, Jacques. 1997. *Fria en Guinée. Des aspects humains d'une industrialisation differente*. Paris: Karthala.

Magrin, Géraud. 2013. *Voyage en Afrique rentière*. Paris: Publications de la Sorbonne.

Mail & Guardian. 2009. Rio Tinto scraps Coega smelter plan on power woes. In *Mail & Guardian*, Oct 16, 2009. http://mg.co.za/article/2009-10-16-rio-tinto-scraps-coega-smelter-plan-on-power-woes. Checked on Sept 30, 2015.

Martin, Lucy. 2015. Taxation, Loss Aversion, and Accountability: Theory and Experimental Evidence for Taxation's Effect on Citizen Behavior. Yale. http://sites.duke.edu/2014bmp/files/2014/10/Martin_TaxAcc.pdf.

Matter, Jean. 1959. Noël 1959. *Bulletin Alucam* (9).

Mina Alumina. 2015. *Our Markets*. http://www.mina-alumina.com/our-markets. Checked on Sept 23, 2015.

Ministère des Mines et de la Géologie. 2013. Titres miniers pour la bauxite. Situation au 31 décembre 2012. Conakry.

Moorcroft, Mia. 2015. Stamp of Approval for Bayside Decommissioning. In zululandobserver.co.za, Aug 15, 2015. http://zululandobserver.co.za/78922/stamp-of-approval-for-bayside-decommissioning/. Checked on Sept 30, 2015.

OEC (2015). The Observatory of Economic Complexity, https://atlas.media.mit.edu/en/. Checked on Sept 23, 2015.

Osborn, Emily. 2009. Casting Aluminium Cooking Pots: Labour, Migration and Artisan Production in West Africa's Informal Sector, 1945–2005. *African Identities* 7 (3).

Otte, Henning. 1998. *VAW versus Pechiney: Les premiers producteurs d'aluminium en France et en Allemagne dans l'age d'or de l'industrie de l'aluminium (1945–1970)*. Aix-en-Provence: Maitrise d'histoire contemporaine.

Oxford Business Group. 2014. *The Report: Ghana 2013*, London. http://www.oxfordbusinessgroup.com/ghana-2013. Checked on Sept 17, 2015.

Pauthier, Céline. 2002. Fria: Une ville-usine en Guinée. Des dernières années de la colonisation à la République Populaire 1956–1973. Mémoire de maitrise d'histoire sous la direction d'Odile Goerg. Paris: Université Denis Diderot Paris VII.

Pechiney Ugine Kuhlmann. 1972. Pechiney Ugine Kuhlmann en Afrique. *Fusion* 5: 2–3.

Pechiney Ugine Kuhlmann. 1979. Les investissements. *Jalons* 24: 9.

Pechiney Ugine Kuhlmann. 1982. Edéa: un cadre de vie agréable. *Jalons* 32: 4.

Pesatimes. 2013. Tanzania—Villagers Blame Bauxite For Endemic Drought. In pesatimes.co.tz, 18/04/2013. http://pesatimes.co.tz/news/energy-mining/villagers-blame-bauxite-for-endemic-drought/tanzania. Checked on Sept 23, 2015.

Pezet, Anne, Marie-Claire Loison, and Céline Berrier. 2009. *Corporate Social Responsibility before CSR. Practices at the Company Aluminium du Cameroun, 1950s–1970s.*

Rainforest Rescue. 2014. *Petition—Stop the Exploitation in the Mountains of Tanzania.* https://www.rainforest-rescue.org/petitions/906/stop-the-exploitation-in-the-mountains-of-tanzania. Checked on Sept 23, 2015.

Rathbone, Richard. 1999. *Nkrumah and the Chiefs. The Politics of Chieftaincy in Ghana 1951–1960.* Oxford: James Currey.

Rio Tinto Alcan (2010): Rio Tinto Alcan Sells Its Share in the Ghana Bauxite Company. http://www.riotintoalcan.com/ENG/media/media_releases_1796.asp. Updated on Feb 3, 2010, checked on Mar 24, 2015.

Riverton, John. 2014. The Volta River Project. In *Global Poverty Alleviation: A Case Book*, ed. P.J. Albert, et al., 31–56. Heidelberg: Springer.

Roberts, Dexter, Chi-Chu Tschang. 2008. Why Chinalco's Buying Into Rio Tinto. *Businessweek*, Mar 5, 2008. http://www.businessweek.com/stories/2008-02-05/why-chinalcos-buying-into-rio-tintobusinessweek-business-news-stock-market-and-financial-advice. Checked on Oct 18, 2013.

Romainville, Michel. 2009. Les routes africaines de l'aluminium. *Techniques & Culture* 51.

Rühl, Lothar. 1959. Frankreichs Zeit in Afrika ist abgelaufen. Ein SPIEGEL-Gespräch mit dem Ministerpräsidenten von Guinea, Sékou Touré. Der Spiegel, Jan 28, 1959, 40–47. http://www.spiegel.de/spiegel/print/d-42624424.html. Checked on Jan 10, 2012.

Rusal. 2013. *History of the Russian Aluminium Industry.* http://www.rusal.ru/en/aluminium/history.aspx. Checked on Feb 26, 2013.

Rusal. 2014a. *Aluminium Producers.* http://www.rusal.ru/en/aluminium/manufacturers.aspx. Updated on Oct 16, 2014, checked on Oct 16, 2014.

Samb, Saliou. 2006. The Electricity Puzzle. *In La Revue - Magazine semestriel du bureau de la banque mondiale en Guinée* 1: 18–21.

Senghaas-Knobloch, Eva. 1975. The Internationalization of Capital and the Process of Underdevelopment. The Case of Black Africa. *Journal of Peace Research 12* (4).

Shoeb, Mohammad. 2012. Production of Aluminium to Grow by 10pc. *The Peninsular*, Nov 22, 2012. http://thepeninsulaqatar.com/

business/215249-production-of-aluminium-to-grow-by-10pc.html. Checked on Oct 10, 2013.

Shundeyev, Vladimir. 1974. The Republic of Guinea. *The African Communist* 58: 34–45.

Soumah, Ibrahima. 2008. *The Future of Mining Industry in Guinea*. Paris: L'Harmattan.

Thaure, Philippe. 2007. *Péchiney vendu: grandeur et décadence du plus grand groupe industriel français*. Paris: Mines Paris.

The Economist. 2015. Wings of Steel. An Alloy of Iron and Aluminium is as Good as Titanium, at a Tenth of the Cost. *The Economist*, Feb 7, 2015.

Touré, Ahmed Sékou. 1967. *L'Afrique et la révolution*. Paris: Présence Africaine.

Track Record. 2010. *Responsible Aluminium Scoping Phase. Main Report*. http://www.trackrecordglobal.com/files/RA%20Scoping%20Phase%20-%20Main%20Report%20v2.pdf. Updated on Feb 28, 2011, checked on Aug 27, 2014.

Udo, Bassey. 2015. Investigation: Inside the Huge Scam Leading to Sale of Nigeria's Aluminium Plant, ALSCON, to Russia's RUSAL. *Premium Times Nigeria*, June 02, 2015. http://www.premiumtimesng.com/investigation-special-reports/184195-investigation-inside-the-huge-scam-leading-to-sale-of-nigerias-aluminium-plant-alscon-to-russias-rusal.html. Checked on Sept 22, 2015.

UN. 2015. Data, http://data.un.org. Checked on Nov 30, 2015.

USGS. 2016. *U.S. Geological Survey*. http://minerals.usgs.gov. Checked on July 15, 2016.

van der Walt, Johannes C. 2009. Desperate Housewives of Empangeni. *The Witness*, 30/01/2009. http://www.witness.co.za/?showcontent&global%5B_id%5D=19043. Checked on May 24, 2013.

Vimetco. 2008. Acquisition of Bauxite Mine in Sierra Leone. http://www.vimetco.com/article/acquisition-bauxite-mine-sierra-leone. Checked on Sept 23, 2015.

Vimetco. 2015. *Our Focus on the Aluminium Industry*. http://www.vimetco.com/about/overview. Checked on Sept 23, 2015.

Vitry, Raoul de. 1957. Éditorial. *Le Bulletin Pechiney* (Christmas issue), 2–3.

Walker, Simon. 2015. Bauxite and Alumina: Growth Maintained. *Engineering and Mining Journal*, 42–47.

Wiederstein, Alexis. 1994. Alu-Wende. Eine weltweite ökologische und soziale Bestandsaufnahme der österreichischen Aluminiumwirtschaft und Aspekte zur Neuorientierung. Wien: Ökologie-Institut.

CHAPTER 2

Concepts: Corporate Decision Spaces and Chains of Command

Until recently, almost all academic studies of the development impact of natural resources focused on the national scale, and were for the most part concerned with the problems and possibilities of resource abundance. The relation between resources and development remains highly disputed, particularly in mainstream economics. After WWII, modernization theorists, drawing upon the notion of comparative advantage, saw mineral endowments as valuable assets for economic take-off, while dependency theorists underlined the structural constraints of mineral production and pointed to declining terms of trade between industrial products and raw materials. The term "resource curse" only became popular in the 1990s (Brunnschweiler and Bulte 2008: 616; Ross 1999: 301) when, for about a decade, the perceptions of most economists seemed to change. This was triggered by the publication of Jeffery Sachs and Andrew Warner's seminal *Natural Resource Abundance and Economic Growth* (1995), which concentrated on the period between 1970 and 1990. The main mechanisms behind what they called a resource "curse" had already been elaborated in the post–war period by the dependency school, whose proponents (1) discovered a long-term fall in prices for raw materials compared to processed goods (the Prebisch–Singer hypothesis), (2) identified high price fluctuations as causes for social and economic instability, and (3) pointed to the emergence of enclave economies and the capabilities of multinationals to repatriate their profits (cf. Ross 1999: 301). These findings became widely accepted in the 1990s and were accompanied by the observation

J. Knierzinger, *Bauxite Mining in Africa*, International Political Economy Series, DOI 10.1007/978-3-319-52706-2_2

that high dependencies on raw material exports spark conflicts and tend to lead to authoritarian regimes (Brunnschweiler and Bulte 2008: 616). To this adds the theorem of the so-called Dutch disease, which suggests that high exports in raw materials lead to stronger currencies and thereby limit the competitiveness of other economic sectors versus foreign firms (Ross 1999: 306).

However, during the commodity boom of the 2000s, even dependency theorists began to question the validity of the theory of declining terms of trade (cf. Jäger and Leubolt 2011: 61). In the same vein, the observation made by Jeffery Sachs and Andrew Warner that extractivist economies perform worse than more diversified economies could hardly be upheld with rocketing resource prices and the relative success of social policies based on resource rents in several "neoextractivist" countries, mainly in Latin America (see Jäger et al. 2014). This change in opinion was also associated with a criticism of the methods that had been employed by the proponents of the resource curse explanation (see Ross 1999: 302–304), which again questioned the causal relationship between resource abundance and poverty. For instance, Brunnschweiler and Bulte point to the fact that resource extraction could be nothing other than a "default sector" in many developing countries, whereby low diversification could simply be induced by weak institutions rather than resource abundance (Brunnschweiler and Bulte 2008: 616). In recent years, with the political consequences of the fall in resource prices—and in particular falling oil prices—in countries, such as Russia, Venezuela, and Brazil, we can finally make out yet another turn in these discussions.

These recurring debates have filled the pages of many a journal and led to the publication of countless policy papers. If anything, it seems to be widely accepted that resource-based development is difficult to achieve in countries with particularly weak political institutions. Yet, even with regard to these special cases, differences of opinion persist over the question of whether these "failures of the state are structural and necessary, or contextual and contingent" (Bridge 2008: 393). Most policy debates tend to depoliticize the discussion by highlighting questions of "state competency and efficiency while obscuring significant divergences of interest between firms, states and consumers that influence the structure of the production network, and its capacity to contribute to the development of regional advantage" (ibid.). This technocratic point of view has been challenged by geopolitical and macro–economic analyses, for instance, under the headline of the "Scramble for

Africa" (see Southall and Melber 2009). However, both the good governance literature and its critics fail to connect their analyses to transnational company strategies as well as to social movements in and around the mines.

2.1 Commodity-Based Approaches

To the geographer John Agnew, one research strategy with which to analyze these entanglements of several scales "would involve tracing the flows of commodities from local African sources to destinations elsewhere, reconstructing the institutional channels and barriers en route" (Agnew and Grant 1996: 742). The most popular framework for this purpose is the global commodity chain (GCC) or global value chain (GVC) (see Bair 2010) and—related to these—the concept of global production networks (Fischer et al. 2010: 14). In the case of aluminum production, the term global production network (GPN) seems the most appropriate. In the following section, I will consider these three closely related approaches as commodity-based approaches. The concept of GCCs was coined in the 1990s by Gary Gereffi and was based upon Immanuel Wallerstein's analysis of "commodity chains", which for him referred "to a network of labour and production processes whose end result is a finished commodity" (Hopkins and Wallerstein 1986: 159).

To date, only a few researchers have employed commodity-based approaches to the extractive sector (see Bridge 2008; Smet and Seiwald 2014; Radhuber 2015; Mayes 2015). This has been for two main reasons: First, it is difficult to describe all the processes that occur in the transformation from raw material to consumer product. Second, material network approaches have been criticized for neglecting the role of state regulation, which is particularly important (or at least more apparent) in the extractive sector. In his initial proposition for doing research on "commodity chains" in the 1980s, Wallerstein explicitly suggested that the idea was to "start with the final production operation and move sequentially backward [...] until one reaches primarily raw material inputs" (Hopkins and Wallerstein 1986: 159). This makes sense given that production can be linked to consumption and thereby to the uneven spatial distribution of workload and its consequences. However, whereas Wallerstein still tried to focus on raw material production as an important node of commodity chains, this was rather neglected in the flood of publications on GCCs in the 1990s, mainly due to the complexity of most commodity chains. As the analysis can only focus on certain aspects of the chain, starting at the "end" usually means sacrificing elements of the "beginning". Raw materials, such as aluminum, cannot be linked as extensively to consumption as the products of the consumer goods industry, the preferred study object of the GCC school.

Moreover, commodity-based approaches tend to over-emphasize the agency of public and private actors in the "Global South" regarding the "up-grading" of commodity or value chains and often leave aside discussions of geopolitical interests and structural considerations (cf. Bair 2010: 24). GVC research, which gained attention a few years after the success of GCC theory, was notorious in this regard as it tended to neglect the questions of power in favor of more policy-oriented aims of inducing growth in Southern economies by linking them up with globalized networks of production. GVC thereby has even made it into policy papers of the World Bank (e.g., Cattaneo et al. 2010; for more information, see Bair 2010: 24, 32–34). This lack of attention to geopolitics and questions of power, as well as a rather technocratic view of the state (as an executor of policies), renders the application of commodity-based approaches on the extractive sector and especially bauxite mining, rather problematic.

Due to the critical importance of bauxite mining and refining for global industries and for the governmental revenues of several producer countries, and due to the oligopolistic structure of the global bauxite/aluminum industry, it is difficult to clearly differentiate between public and private actors, both in terms of organizational structure and in terms of rationalities of firms and governments. Global players in the aluminum sector, such as Alcoa, Rusal, or Rio Tinto, often operate beyond the conditions of free-market competition and are in many instances too "big to fail" (cf. Taylor 2000: 7f). At the same time, the governments of resource-rich countries periodically act like private shareholders (cf. Shafer 1986: 925). Most major companies are strongly connected and supported by their states of origin (see American postwar production, Pechiney in the 1980s or Rusal in the 2000s). The governments of the African bauxite producer countries are shareholders in mining joint ventures, providing large parts of government revenues. From independence until the recent resource boom that led to a rise of gold exports, bauxite has been responsible for 80–90% of exports of Guinea and thereby for most government revenue (cf. EITI 2015). The exports of Sierra Leone consist mostly of iron ore and Ghana mainly exports gold, cocoa beans, and lately also crude oil. Historical price fluctuations of these and other minerals led to periodical budget crises, forcing the governments to cut public employment and social spending and leading to dependence on aid and foreign loans.

In politically unstable regions, the few globally dominating aluminum companies (the "majors") rarely own one mine alone. They share these mines with their 'competitors' and the involved states, in order to reduce the risks of these huge and long-term investments. These risks include expropriation, price fluctuations, social upheavals, war, etc. In addition, the aluminum sector is very capital-intensive and highly concentrated (cf. Campbell 1986: 2, 5). As subsidiaries of vertically integrated corporations, bauxite mining joint ventures function very differently from normal firms, given that they are not independent profit-maximizing entities. Their role is to produce raw materials at a low cost, although very often not in competition with other subsidiaries in the same country or mine. The so-called efficiency of the market only comes into play at much higher tiers when the mother companies sell their aluminum ingots or even semi-processed goods, such as parts of airplanes. The small share of the value added at the bauxite mining stage thus rather leads to rationalities of centrally planned economies: The mines primarily have to reach certain output goals and serve as a link between the parent group and the government (cf. Habig 1983: 247–251; 262–263). Prices are mostly set by pre-arranged formulas (Compagnie des Bauxites de Guinée 1983: 14). In times of price declines—for instance, in the early-1990s and again after 2007—the dominating companies coordinated their reduction of output (see Larrue 1997: 300; Deripaska 2014). Consequently, bauxite prices are only weakly correlated with world aluminum prices. For instance, while aluminum prices doubled at the end of the 1980s, the bauxite price of the *Compagnie des Bauxites de Guinée*—the largest Guinean producer—fell from 35 to 25 USD (cf. International Monetary Fund 2008: 7). The resulting chains of command are thereby a mixture of public and private structures, with the latter being much more interrelated and much less exposed to competition than capital in most other sectors. A meaningful analysis of African bauxite production thus has to address (1) the interests of producer countries, receiver countries, and international institutions, (2) risk management and profit-maximizing strategies (in other words, chain governance) of TNCs, and (3) the renegotiation and the effects of this global production network at the local level.

2.2 GCC, GVC, or GPN?

Due to this need for a broad analytical framework and due to the methodological challenge of effectively connecting raw material production with consumption, only a few studies on raw materials have used these network approaches to date. Interestingly, all of these studies use the concept of global production network, even though the differences of the three presented concepts—GCC, GVC and GPN—are rather negligible.[1] Whereas GPN theorists have reacted to the criticism of neglecting regulation by underlining the importance of the "embeddedness" of production networks, the sociologist Jennifer Bair proposes a "second generation" of GCC research, focusing more on political economy and returning to the analysis of power relations that has been neglected with the advent of GVC analysis. Bair criticizes neoliberal applications of the theory (under the title "up-grading"; see Bair 2010: 30), pleads for a stronger consideration of working conditions and reproduction, and suggests linking the GCC concepts with structural analysis of capitalist production (Bair 2010). My choice to use the label of global production networks has, therefore, rather pragmatic reasons beneath it: Connections between GPN and the extractive sector have already been established by several studies, and GPN is the last offspring and thereby the most topical label of commodity-based approaches.

2.3 Possible Contributions of the Extractive Sector to GPN Research

Working with GPN allows overcoming the trap of "methodological nationalism" (Brenner 1999; Jessop et al. 2008), which has largely contributed to an "analytical stalemate" concerning the resource curse thesis in the last few years (Bridge 2008: 390). Moreover, several discussions around GPN can be enriched with empirical findings from mines and drilling fields. For instance, in most of the described cases of commodity-based research, power is exercised against the flow of commodity production. The higher the value added of a product in a certain chain, the more power its holder can exercise throughout the chain. Many IT and apparel branded companies have outsourced almost all of their production capacities in the last 30 years, albeit retaining quite a firm control over the whole commodity chain (Plank and Staritz 2010). However, in the case of metal commodity chains—and especially in the case of the aluminum commodity chain—large parts of the downstream sectors are actually controlled by the mining companies. This was the case in the USA until the 1960s and can still be seen for Russian and Chinese aluminum companies. In Germany, the dominant industrial companies recently formed an alliance in an attempt to regain control over raw material producers (Allianz zur Rohstoffsicherung; cf. Heß 2012). The result of these permanent tensions over chain governance is a multipolar value chain, something that has not been theoretically considered to date (Raikes et al. 2000: 7). This is particularly interesting if combined with the long-term development of raw material prices: In the 1970s, when raw material prices were almost as high as they are nowadays, raw material producers (countries, regions, companies, etc.) gained as much bargaining power as they did recently during the resource boom in 2004–2008.

Second, through its connections to Wallerstein's world system theory, the categories of the GPN school can easily be related to dependency theory which had largely been neglected in the last 30 years and has recently received renewed interest (Pimmer and Schmidt 2015). GPN concepts are well-suited tools with which to analyze power differentials between decision spaces. They enable "zooming" into concrete power relations without losing the overall politico-economic context and thus are part of the exclusive club of sociological middle-range theories that can connect micro- and macro-analysis. As I will show based upon the concrete example of Guinean mining towns, GPN concepts also allow for a sociological reformulation of Wallerstein's idea of a zero-sum game of development. By introducing the term "global chains of command", I would like to show that control over people can be described as a zero-sum game; however, contrary to Wallerstein's concept, this control does not automatically imply a zero-sum game of development.

During my stays in Guinea, I was confronted with a very practice-oriented version of GVC that had been tailored to the analysis of mining enterprises by public administrators and politicians. This was particularly astonishing as GVC theory has rarely focused on mining to date. This theory was presented during a workshop for "civil society activists" on the new mining code in Kindia, which was organized and sponsored by the Revenue Watch Institute (Revenue Watch Institute 2014: 2; Legal expert 13/03/2014). The RWI recently changed its name to Natural Resource Governance Institute (NRGI) and is mainly financed by George Soros' Open Society Foundations (see below). On its homepage, the institute provides an account of this extractive industry value chain approach:

"In his book The Bottom Billion, Paul Collier popularized this approach to stress the key steps in ensuring that natural resource wealth transforms into citizen wellbeing. This framework has since become a reference for other organizations working on natural resource governance, such as Revenue Watch Institute, the World Bank, and the Extractive Industries Transparency Initiative (EITI)" (Revenue Watch Resource Center 2015).

On the same page, the NRGI concedes that there are other forms of value chain analysis that more strongly focus on the private sector. By contrast, the approach presented by the NRGI "takes a government perspective to optimize value for a country and its citizens" (ibid). Paul Collier divides the extractives value chain into five steps: "Deciding to Extract", "Getting a Good Deal", "Ensuring Revenue Transparency", "Managing Volatile Resources", and "Investing for Sustainable Development" (ibid). Therefore, Collier's "approach"—as presented on this homepage—does not have much in common with GVC theory: if anything, it could be called a policy application of the concept. Nonetheless, its focus on governmental revenues makes sense given the low labor intensity of the mining industry. Contrary to the processing industry, tax revenues are considered among the most important contributions of the mining industry to economic development (cf. UNCTAD 2007: XXIV; see also Open Society Institute of Southern Africa et al. 2009: 2). During the workshop in Kindia, the instructors also presented a "Chain of Change" (*Chaîne de changement*) with some more steps to follow that has been applied since 2012 (Interview with Legal expert 11/03/2014). According to several interview partners in Kindia, Paul Collier is a close advisor to Alpha Condé, the current president of Guinea (Interview with Legal expert 11/03/2014; see also Collier 2007; Taylor 2011).

2.3.1 *Corporate Decision Space and Infrastructural Power*

Up to a certain degree, the material flows of aluminum are framed by (classical) state-centered regulation. Bauxite produced in the mines of Guinea transcends borders—to be refined, smelted, processed, and consumed—by moving according to "traffic rules" of national and international regulation. Given that many of these "traffic rules" are not standardized on a global level, this framing of material flows often leads to conflicts over regulations. One of these conflicts of rules was China's decision to restrict its raw material exports in the early 2000s, which also concerned bauxite, and heavily contributed to the recent price boom in raw materials from 2004 to 2008. The national regulation of China—which has been in conflict with a certain interpretation of the WTO treaties—restrained aluminum atoms from transcending Chinese borders until they had been processed into consumer products. These zones, where legislations overlap or are in conflict with one another, offer particularly ample room for maneuver. Actors have the possibility to juggle with regulations by changing sides or jumping scales (cf. Jessop et al. 2008: 390; Dicken et al. 2001: 95), and they have an increasing influence on the emergence of new orders that could be established within or above the existing ones. If economic actors thereby transcend political regulations or boundaries, they mainly do so by building upon other, conflicting regulations or boundaries often coming from other scales (e.g., national versus international regulations).

West African mining towns are special examples of such zones of superposition, given that they contain a great number of conflicting orders, many of which are marginally formalized. The political structure of the towns already depends on conflicting interpretations of different generations of national law, additionally being in conflict with (neo-) traditional orders and interpretations of the customary law. Bierschenk and Olivier de Sardan called this sedimentation of political institutions of several epochs "institutional addition" (Bierschenk and Olivier de Sardan 1999: 52). This leads to a "flexibility and malleability in these institutions [...], leaving ample room for negotiation, usually informal, for the different actors who must define required competences and rules of the game" (Bierschenk and Olivier de Sardan 2002: 4; cf. Knierzinger 2011: 33).

This situation is additionally complicated by yet another form of reference that is neither part of the national framework of the formal political system, nor of (neo-)traditional ones. This reference is the transnational decision space[2] of mining companies, which is dependent on international agreements (e.g., the IFC Performance Standards on Environmental and Social Sustainability; International Finance Corporation 2012), company codes and standards, national law and governmental decisions of their countries of origin, and—much more than most other spaces—personal decisions of several individuals that can only be culturally and socially defined. In remote places, companies often dispose over an almost physical sort of power based upon infrastructure, which creates considerable room for maneuver that is neither codified by written texts like national laws nor codified by company regulations, leading to some sort of corporate neopatrimonialism.

Drawing upon Agnew's reflections on power, I would like to call the basis of this special form of control "infrastructural power". Agnew borrows the concept from the sociologist Michael Mann, who differentiates between despotic power—which "denotes power by the state elite over civil society"—and infrastructural power—which "denotes the power of the state to penetrate and centrally co-ordinate the activities of civil society through its own infrastructure"—as the main forms of powers of states (Mann 1984: 188; cf. Agnew 2005: 443). For Agnew, infrastructural power can be derived from roads, electricity and tap water, as well as deterritorialized infrastructure, such as "currencies, systems of measure, trading networks, educational provision [or] welfare services" (ibid).[3] As will be shown in Chap. 3, the Guinean government almost exclusively bases its control over the population on despotic power in the sense of Mann, whereby infrastructure is mostly controlled from the outside. Besides international financial institutions, development agencies and the African diaspora (in the form of remittances and investments) mining companies play a central role in providing this infrastructure.

Having been confronted with a lack of infrastructure, mining companies were forced to build and maintain whole cities and railways in Guinea. Directors of mines thus control most of the city's buildings, its electricity stations, health system, waste management, etc. Similar to the strongly personalized political systems in West Africa and comparable to the position of the colonial governor, the directors of the mining towns are thereby confronted with relatively few checks and balances and exert considerable control over these cities. However, through the corporate chain of command, several other corporate actors also become political entrepreneurs. From a historical perspective, West African trade unions play a particularly important role as links between private governance and formal government. In Guinea, trade unions were central to both the independence movement and subsequent state formation processes and they continue to actively participate in Guinean politics.

2.4 Global Chains of Command

As a consequence, the decision space of companies almost entirely replaces the formal political space in remote West African mining towns. Due to their economic position as export enclaves, these cities seem to be more strongly interlinked with the corporate governance structure of the mine operators than with the concentric political hierarchy of spaces in the area. Democratic participation—which also remains quite a theoretical concept in most other parts of the world—is thereby replaced by the decisions of one or several companies who control the infrastructure, the development of the local economy, and in several instances even possess the monopoly of violence. The already fragile democratic system in numerous African countries thereby overlaps with a system of company rule, which can only be called despotic as it solely relies on a hierarchical chain of command with global investors at the top. Whereas most of the workers and the surrounding population in mining towns of industrialized countries have the option to "exit" (quit the job) or "voice" (protest or appeal), these options exist to a far lesser extent for most Africans.

The term "chain of command" is a concept that enables linking these discussions on governance with GPN theory. Before focusing on "upgrading", Gary Gereffi used the concepts of GCC mainly to explain power relations in global production processes. One of his most important points was the insight that companies originating from the old industrial centers were still in control of global production without actually employing large numbers of workers anymore. Rather than explaining this paradox as a temporal phenomenon, Gereffi introduced the concept of buyer-driven commodity chains. These chains are "steered" by brand companies who mainly profit from their control over consumer markets (Gereffi 1997: 116f).

While Gereffi uses the term "drivenness" to explain this form of control, I would like to formulate this power relation more concretely by talking about global chains of command. In classical management theory, a chain of command is a simple hierarchy of orders from the CEO over various forms of supervisors and foreman to a blue-collar worker, provided that such a worker exists in the company. The term was coined by Henry Fayol (1929), director of the *Société de Commentry, Fourchambault et Decazeville,* a large metallurgical company founded in the mid-nineteenth century. Today, most of these chains of command stretch across the whole globe and can become effective within one single company—this would be a producer-driven commodity chain in Gereffi's terminology—or be split up into several companies, whose only link is the market economy—this would be the buyer-driven commodity chain.

Using the concept of global chains of command, we can also incorporate phenomena such as unpaid (household) labor, changing local power hierarchies, the social situation in African mining towns, and even national political developments, into the analysis of the GPN of aluminum. For instance, when the CEO of Rusal decided to close the factory in Fria in 2012, he thereby stopped cash flows that hierarchically structured a whole region and even threatened the national political establishment. The spreading misery (from workers' families to merchants, artisans, and farmers) thereby went hand in hand with a partial reversal of power relations: now, women increasingly replaced the predominantly male workers as negotiators and demonstrators (cf. Rubbers 2010: 336); relatives from the countryside who had before sent their children to Fria to live with the workers' families now received back both their own and the children of the workers; retailers who had previously lived from the worker's salaries now became donators of rice; and local politicians partly re-established control over the city. Rusal thereby interfered into previously existing hierarchies in Fria that did not seem to be connected to economic control relations, particularly from a Western perspective.

The interaction between unequally powerful and unevenly trained actors can have positive long-term effects, leading to accelerated development through the diffusion and the exchange of knowledge. However, the economic exchange between unequal partners also has an almost inevitable short-term consequence, which can be termed exploitation. The world economy increasingly connects people with very different degrees of social protection and economic capital, leading to an increasingly uneven exchange of working time. For instance, unskilled dock workers in Kamsar only have a fraction of the purchasing power of unskilled dock workers in Hamburg, despite a good number of them doing the same work for the same TNCs. In the case of the relations between capital and labor, this uneven relation of control reaches a level that can clearly be called "political": the major investors of the aluminum companies have incomes that match the aggregated income of tens to hundreds of thousands of Africans. Whereas the current CEO-to-worker compensation ratio in the USA is around 500:1 (compared to 30:1 in the late 1970s; see Mishel and Davis 2014), the income relation between the majority owner of Rusal and the average income of petty traders in Fria probably reaches a rate of 100,000:1 (cf. Knierzinger 2014b). This huge income gap leaves its marks. In Mambia, for instance (see Sect. 3.1), at the same time when the population had taken to the streets, because Rusal had not paid its taxes, the sub-prefect had his residence refurnished by the same company (Sous-Préfet of Mambia 18/03/2014). The current mayor of Kamsar—who was made responsible for considerable destruction during demonstrations—was sent to Mecca by CBG for more than a month in 2013 despite having already left the company 8 years before. The reason for this present was anything but a secret (cf. Executive CBG 17/02/2014; Vice mayor of Kamsar 19/02/2014). The police commissioner of Kamsar stated that his mission "here is the CBG". He lives in a house built and maintained by CBG, has access to the company's cheap food supplies from abroad and receives 40 L of fuel per month, a practice that has no legal basis and is often denounced as corruption(Interview with Police Commissioner 17/02/2014).

However, it is important to note that most of these practices do not violate any law and thereby cannot be called corruption. They are rather wide-ranging (and yet academically widely unnoticed) political side effects of capitalist "Landnahme" ("taking of the land"; Dörre 2010) in countries like Guinea. The proposed notion of global chains of command aims at these very concrete relations of control, based, on the one hand, upon the transnational company hierarchy and, on the other hand, on the simple fact, that having reached a certain level of income, money is primarily a means of power. Once it cannot be spent to tell others what they should produce or do for us (consumption), it only serves to tell others what they should do for others (investment). Thus, from this sociological point of view, investment is a purely means of command similar to political rule.

2.5 Crises of Chain Governance

It is important to note that the power relations within this global network of public and private actors change across time. This historic dimension of (corporate) governance has not received much attention to date (cf. Hönke 2009: 7; Bridge 2008: 411, 414). To capture this dimension, I identified two crises of global raw material governance that held particular importance to the bauxite production network in Africa but could also be applied to various other countries and sectors (Knierzinger 2014a: 21; cf. Southall, Melber 2009: XIX). These two crises happened in the 1970s and in the mid-2000s and were triggered by raw material and energy shortages in the old industrialized centers. Both led to a steep rise of raw material prices and thereby prompted resource-rich governments to (re-)gain control by forming alliances, introducing new legislations and increasing the mining assets of the state, which again prompted reactions by industrialized states and TNCs. These general similarities between the 1970s and the 2000s can also be found in the global aluminum industry: when resource-rich governments started to nationalize and raise taxes in the 1960s and formed the International Bauxite Association (IBA) in 1974[4]—explicitly modeled on the example of OPEC—both the aluminum oligopoly of that time (the Six Sisters) and the governments of the consumer countries reacted by redeploying production (OCED countries obtained their bauxite increasingly from Australia) and trying to co-opt the "insurgents". For instance, this involved joint ventures with governments in Africa and the Caribbean and new aid agreements (see Table 2.1). In most of the concerned countries, these struggles were additionally accompanied (and triggered) by declarations of independence, the resulting non-aligned movement and the new social and solidarity movements in the consumer countries that were highly entangled with the green movement. I have tried to summarize the successive steps of these crises of raw material governance in Table 2.1.

These two crises of chain governance went along with shifts in the repartition of the command over global working time. The significant rise of living standards in Arab countries, which was similar to the considerable social mobility witnessed in South America and China during the last commodity boom, went along with higher costs of living (and thereby less control over global working time) in the old consumer countries. Similarly, the elite of producer countries took over a significant part of the control over workers of mines, factories and supermarkets from the old industrial centers (cf. Bridge 2008: 402; Bush 2010: 241).

The place of Sub-Saharan Africa in this political process remains contested: it is often considered to have been left out of globalization, simply because its share in world trade has diminished or stagnated since the 1970s while other regions have become increasingly interdependent (e.g., Krause 2002: 24; see also Agnew and Grant 1996: 729–730).[5] However, the figures showing this lack of exchange do not capture the actual number of Africans connected to the world, because they do not capture the development of unequal exchange: If average Africans receive increasingly less for the same amount of work—as was the case from the

Table 2.1 Two crises of chain governance

	1960s/70s: Crisis of decolonization	**2000s: New Scramble for Africa**
Causes	**Independences, Yom Kippur war, Europe & Japan↑**	**Financial crisis, BRICS↑**
(Re-)actions of Producer Countries	Political East-West-oscillation	BRICS as new investors/donors
	Nationalizations	Public private partnership
	Higher royalties, export restrictions, etc.	
	Producer organizations[1]	
(Re-)actions of civil society	New social movements (Environmentalism, strikes, etc.)	Alter-globalization movement and strikes
Effects	**Oil prices↑ raw material prices↑**	
Reactions of TNCs	More exploration & mining in OECD countries	
	Joint Ventures, vertical integration	
	Deconcentration: Nationalization	Rather Concentration: TNCs and China
Reactions of OECD countries	Stabex & Sysmin (Lomé I&II)	Transparency initiatives: EITI, Dodd-Frank, etc.
	Pressure for free markets: Treaty of Rome, Yaoundé I-II, Lomé I-IV	Bi– and multilateral trade agreements[2]
	Mounting East-West tensions	Increased military presence in Africa

[1] E.g. Organization of the Petroleum Exporting Countries, International Bauxite Association and Intergovernmental Council of Copper Exporting Countries.
[2] Economic Partnership Agreements

1970s until the turn of the millennium—their share in world trade evidently falls. Since the turn of the millennium, with the new price peak in raw materials, a veritable euphoria has taken place as Africans have seemed to increasingly participate in world trade. However, in many raw material sectors, only the yields per unit—as opposed to the number of employees—have significantly increased. With falling raw material prices owing to the cooling of the world economy, African elites are about to lose these windfall profits again.

The recent shift of power in the 2000s led to numerous promises from mining companies to process bauxite in the producer countries, quite similar to promises made in the 1960s and 1970s. In both cases, with falling prices, most of these plans vanished into their computer folders again. In Guinea, as in many other resource-rich countries, this also concerned the legal framework of mining: In 2011, the recent crisis of raw material governance resulted in a new progressive mining code. However, only 2 years later, due to intensified international pressure, the government felt obliged to revise this code again, mainly by reducing taxation and softening regulations concerning state participation (for more theoretical discussion, see Sects. 2.2 and 2.3).[6]

One important feature of both crises is their profoundly political nature. Similar to the rise of oil prices owing to the Yom Kippur War in 1973, the recent fall of oil prices is much more a consequence of political decisions by concrete actors than the expression of a mechanical world economic cycle of boom and bust. One of the most important decisions in this respect was that of the Chinese government to restrict the exports of minerals, such as zinc, bauxite, magnesium, phosphor, silicon carbide, silicon metal (silicon ore), coke, and fluorite through export taxes, quotas, minimum prices, and additional requirements for foreign companies (Kauder et al. 2011: 2). In 2011, as a consequence of a conflict over the Senaku islands, the committee also banned the export of rare earths to Japan (Kauder et al. 2011: 2).

In 2009, the USA requested consultations with China within the conflict resettlement body of the WTO with respect to China's export restrictions. Four years later, on 28 January 2013, the Chinese government finally met the recommendations made by the WTO and lifted the export restrictions. During the process, China had mainly argued that the restrictions were (1) necessary for the protection of the health of its citizens and (2) related to the conservation of exhaustible natural resources. In the case of bauxite, the second argument was used: An export quota on refractory-grade bauxite was "temporarily applied" to either prevent or relieve a "critical shortage". Both arguments were dismissed by the dispute settlement body (WTO 2013). Due to the unclear clauses in the WTO concerning export restrictions, many developing countries used these instruments during this crisis of the 2000s, not least because they were among the last instruments of that sort (cf. Lambert 2012: 132).

It is also interesting to note that the media coverage of both crises—which in turn potentiated the resource power of resource-rich countries—was mainly driven by mining and processing companies themselves. At the World Economic Forum in 2007, the heads of over a dozen Western mining companies—including Rio Tinto, Anglo American, and Newmont—complained that they were increasingly being cut out of deals by state-owned Chinese companies and appealed to the UN and the World Bank to insist upon higher environmental and safety standards while maintaining "that Africa was being raped and pillaged by China" (Robertson 2007; cf. Southall 2009: 16). The companies' imploring demands for public guarantees, funds, and political support to secure their supply chains also led to the opposite: Reassured of their political gains and resource-rich governments intensified their cooperation.

The spectacular rise of Chinese industrial production obstructs the view on other semi-industrialized countries, such as India, South Africa, and other populous countries that were part of these developments in terms of both economic development and related political decisions. In the case of Indonesia, this "revolt" of the resource rich "Global South" also backfired on China when the Indonesian government quite resolutely changed course, aiming at integrated production, higher taxes, and better social and ecological standards (Nodalis et al. 2013: 14). In June 2012, as a consequence of export restrictions, China's bauxite imports from Indonesia declined from 80% of its overall bauxite imports to almost zero within 1 month. According to media reports, Chinese companies reacted by trying even harder to gain access to Guinea which did not manage to implement such restrictions (Bah 2014; Oumar 2014).

Another important feature of both crises is attempts of the dominant players to co-opt other actors within their global production networks. Similar to the aid agreements of Lomé, the recent raw material boom went along with various attempts to reduce the negative impact of mining in the Global South. This concerns several attempts to regulate the markets for the so-called conflict minerals, such as the Kimberly process starting in 2003, parts of the Dodd-Frank Act (1502) that passed the US parliament in 2010, and a similar initiative of the EU (European Commission 2014). In contrast to the Dodd-Frank Act 1502, the latter only proposes voluntary measures and has been heavily criticized by NGOs (Küblböck 2014; cf. Global Policy Forum Europe 2011: 8).

Apart from that, governments, IFIs, and international NGOs also pushed for more transparency in the whole sector. Section 1504 of the Dodd-Frank Act would force all companies of the extractive industry listed on stock exchanges in the USA to publish payments to all governments where they do business. However, in October 2012, the American Petroleum Institute, the US Chamber of Commerce, and other institutions successfully appealed against the law. By the time of writing, no amendments had been enacted. With the Accounting and Transparency Directives of 2013, the EU has created similar conditions for companies in Europe, but also in this case, the respective amendments are still to be implemented at the national level at the time of writing.[7]

With the Extractive Industries Transparency Initiative (EITI), launched at the Earth Summit in Johannesburg in September 2002 on the initiative of Tony Blair, similar measures have already been taken in over 50 member countries, mostly in the Global South. The Extractive Industries Transparency Initiative (EITI) is a global Standard that promotes open and accountable management of natural resources, mainly by comparing the declarations of mining companies and producer countries concerning the payment of taxes and royalties. All African bauxite producer countries as well as most of the aluminum majors are part of the EITI (EITI 2015). In Guinea, as in many other institutionally weak countries, it was implemented with the help of the "Revenue Watch Institute" (RWI) and Publish What You Pay (PWYP). RWI is a "non-profit policy institute and grantmaking organization" initially founded "through the generous support of the Bill and Melinda Gates Foundation, the William and Flora Hewlett Foundation, and the Open Society Institute, along with several other donors" (Natural Resource Governance Institute 2013). PWYP is a "global network of civil society organizations united in their call for an open and accountable extractive sector" and also receives the majority of its funds from the Open Society Foundations (PWYP 2015). The latter is established and financed by George Soros.

In addition, the EITI was heavily supported by the German GIZ under the headline "Good Governance" (Interview with Consultant 10/02/2012). The central tool of the EITI is the annual publication of payments from mining companies to governments. It is not voluntary as it primarily addresses governments and encourages them to force companies to disclose their payments, but the enforcement mechanisms are weak and the publication most of the time takes place only in an agglomerated version of all disclosed payments in a country, which means one figure for all company payments being compared with the figures of the government (see Moran 2013). Nonetheless, the citizens of these states now not only have the possibility of questioning the use of funds that were previously not even known, but they also gain a clearer picture of the scope of the government's dependency on mineral resources. According to the EITI website, government revenues of Guinea from the six mining companies reporting at present[8] rose from 140 to about 220 million USD from 2005 to 2010 and mining remains responsible for about 95% of exports (EITI 2015).

While about two-thirds of all EITI countries are situated on the African continent, Latin America only counts two (Guatemala and Peru) and the industrialized and industrializing countries are almost completely missing; neither Australia nor any of the BRICS countries have produced EITI reports to date (EITI 2015). Even the countries that have pushed the agenda from the beginning, the USA and the EU have not yet implemented similar legislation. Section 1504 of the Dodd-Frank Act in the USA and similar legislation in the EU would theoretically go further than the EITI, but none of these provisions has to date been implemented.

2.6 The African State as a Victim of Globalization?

The media and academic publications often implicitly depict politics or the state as a victim of corporate globalization, despite the fact that the opposite—namely, a clear primacy of politics over economy—would neither automatically lead to better living conditions nor to less exploitative systems. Fordism has often been described as a positive counterexample in this respect, as capital mobility was relatively low and thus allowed for a series of political concessions toward the working force, especially vis-à-vis its male part (Becker 2003: 71). This perspective is quite useful when applied to the European or North American context, particularly because the call for the democratization of the economic sphere can be derived from this assumption, but it also leads to a simplified understanding of politics, which is particularly problematic in post-colonial contexts.

The normative connotation of the need to enclose economic flows changes the meaning of both economic flows and political control.

Economic flows appear as the opposite of political control and thus as irrational, apolitical, and quasi-natural processes, which seem to be located outside of (conscious) social actions. Depicting capitalism as wild and unregulated naturalizes its numerous institutional bases, such as property, the acceptance of currency systems, the pursuit of profit, etc. Many scholars try to evade this normative connotation by explicitly dismissing both the assumption of a complete domination of the state by capital and the idea of the state as an instrument of "the people". Recently, this has often been explained by drawing upon the work of the Greek-French scholar Nicos Poulantzas who warned of "authoritarian statism" as the "new" type of the capitalist state in the 1970s (e.g., Kannankulam 2006). To avoid describing the state as the savior of humanity against unchained economic forces, this academic tradition defines the state as the most important lieu of temporal fixation of antagonist (class) interests (e.g., Brand et al. 2007: 222; Harvey 2006). For me, this is also the only definition of the state that enables a common designation of its quite different formations in (semi-)industrialized countries and non-industrialized countries, such as Guinea. During the twentieth century, a variety of discursive practices led to the impression that the world is made up of a certain number of identically structured and identically functioning nation-states (cf. Agnew 2005: 440). This pitfall can hopefully be avoided by defining states as the most important nodes of global power relations, bearing in mind that the size and the importance of these nodes vary considerably.

With the rising importance of corporate social responsibility (CSR) in extractive industries in recent years, these discussions about state regulation in a globalized economy were subjected to a qualitative shift. While until recently, social services and the protection of the environment had been to a certain degree a necessary evil to mining companies, the same actors now embrace these measures as means of marketing and generators of legitimacy.[9] This means that while economic space has always simultaneously been political space, since the 1980s, it also seems to have become increasingly discursively constructed and legitimized as such. Under the title of "politics of discharge" (cf. Hibou 1999), several recent studies on mining in Africa nonetheless came to the conclusion that international mining companies do not replace the state but rather enforce it (e.g., Williams 2010: 128; Hönke 2009: 14). I would also challenge the idea that from a global perspective, states and governments have lost power in comparison with transnational corporations during the last 30 years. However, liberalization nonetheless seems to be an instrument of choice of "chain commanders" (public and private, i.e., Northern governments and transnational investors) to (re-)assert their position in their global chains of command. I, therefore, doubt the underlying assumption of Williams, Hönke, and Hibou that the interests of most multinational corporations and African political elites tend to converge in the long run. It is true that privately-led mining endeavors have filled the state treasuries since independence, thereby laying the basis for a stronger state; however, many African regimes have pursued goals that were incompatible with the profit-maximizing interests of the international mining companies and have only forcibly liberalized during the debt crisis of the 1980s. Chapter 3 will show that these fragmented constellations of the rule also prevented effective state formation in the long run and have, therefore, effectively weakened African governments.

Notes

1. "Despite the concerted efforts of its proponents to distinguish the GPN framework from the GCC approach, the research carried out under the banner of the former does not differ greatly from studies of global commodity chains, particularly in terms of methodology. [...] While the theoretical elaboration of the GPN approach calls attention to aspects of the GCC framework that have been largely neglected in analyses of global commodity chains, the GPN research agenda does not differ substantially from empirical work in the GCC tradition" (Bair 2008: 356). Describing the even less important difference between GCC and GVC, Bair writes of a "mere playing upon words" (Bair 2010: 32).
2. I borrowed the term decision space from Charles Maier's differenciation between decision space and identity space. He connects the increasing separation of these two spaces with the decling importance of "territoriality" (Maier 2006: 49–51).
3. This infrastructural power lays the basis for the ascribed power of what he calls a "sovereignty regime" (see above).
4. See Holloway (1988: 44–55), Forster (1976: 135), Habig (1983: 136, 205), cf. Soumah (2008: 170).
5. Agnew also argues against the thesis of Africa "falling out" of world trade, albeit not on the same basis: he does not criticize the instruments of measure but rather the selection of countries.
6. Cf. Norton Rose Fulbright (2013), République de Guinée (2011, 2013), Revenue Watch Institute (2013), Wann (2014).

7. Küblböck, Pinter (2015), European Commission (2013), see also Global Policy Forum Europe (2011: 9), Jaeger (2012: 2), Kerkow et al. (2012: 21), Legal expert 11/03/2014.
8. In 2010, seven companies reported. For 2011 and 2012, EITI Guinea states that is still in the process of hiring the "reconciler", a professional services firm that compares the declarations of the government with those of the companies. Until now, this was conducted by KPMG, Deloitte and Moore Stephens LLP (EITI 2015).
9. For the aluminum industry, see e.g. Aluminium Stewardship Initiative 2015; cf. Hönke (2009: 2–4), Pezet et al. (2009), Strange (1996: 44).

References

Agnew, John. 2005. Sovereignty Regimes: Territoriality and State Authority in Contemporary World Politics. *Annals of the Association of American Geographers* 2 (95): 437–461.

Agnew, John, and Richard Grant. 1996. Representing Africa: The Geography of Africa in World Trade, 1960–1992. *Annals of the Association of American Geographers* 86 (4): 729–744.

Bah, Boubacar Caba. 2014. Mines: Hongqiao Group, le mastodonte chinois de l'aluminium rachète AMC Bauxite pour 120 millions USD. In http://guineenews.org, 9 June 2014, http://guineenews.org/mines-hongqiao-group-le-mastodonte-chinois-de-laluminium-rachete-amc-bauxite-pour-120-millions-usd/, checked on 14 Jan 2015.

Bair, Jennifer. 2008. Analysing Global Economic Organization: Embedded Networks and Global Chains Compared. *Economy and Society* 37 (3): 339–364.

Bair, Jennifer. 2010. Globaler Kapitalismus und Güterketten. Rückblick und Ausblick. In *Global Güterketten—Weltweite Arbeitsteilung und ungleiche Entwicklung*, ed. Karin Fischer, Christian Reiner, and Cornelia Staritz, 24–43. Wien: Promedia.

Becker, Joachim. 2003. Beiderseits des Rheins. Regulationstheorie und emanzipatorische Politik. In *Fit für den Postfordismus? Theoretisch-politische Perspektiven des Regulationsansatzes*, ed. Ulrich Brand, and Werner Raza, 58–76. Münster: Westfälisches Dampfboot.

Bierschenk, Thomas, Olivier de Sardan, and Jean-Pierre. 1999. Dezentralisierung und lokale Demokratie Macht und Politik ländlichen Benin in den 1980er Jahren. In *Dezentralisierung, Demokratisierung und die lokale Repräsentation des Staates. Theoretische Kontroversen und empirische Forschungen*, ed. Trutz von Trotha, and Jakob Rösel, 37–68. Köln.

Bierschenk, Thomas, Olivier de Sardan, and Jean Pierre. 2002. Local Development Brokers in Africa. The Rise of a New Social Category.

Department of Anthropology and African Studies Working Papers Vol. 13, Johannes Gutenberg University, Mainz.

Brand, Ulrich, Görg Christoph, and Wissen Markus. 2007. Verdichtungen zweiter Ordnung. Die Internationalisierung des Staates aus einer neo-poulantzianischen Perspektive. *PROKLA. Zeitschrift für kritische Sozialwissenschaft* 147 (2): 217–234.

Brenner, Neil. 1999. Beyond State-Centrism? Space, Territoriality, and Geographical Scale in Globalization Studies. *Theory and Society* 28 (1): 39–78.

Bridge, Gavin. 2008. Global Production Networks and the Extractive Sector: Governing Resource-Based Development. *Journal of Economic Geography* 8: 389–419.

Brunnschweiler, Christa, and Erwin Bulte. 2008. Linking Natural Resources to Slow Growth and More Conflict. *Science* 320: 616–617.

Bush, Ray. 2010. Conclusions: Mining, Dispossession, and Transformation in Africa. In *Zambia, Mining, and Neoliberalism*, ed. A. Fraser, and M. Larmer, 237–268. New York: Palgrave Macmillan.

Campbell, Bonnie K. 1986. *Bauxite Bargaining. The Consequences for a Bauxite-Producing Country of the Renegotiation of the Bauxite Levy: The Example of Guinea*. Vienna: United Nations Industrial Development Organization.

Cattaneo, Olivier, Gereffi Gary, and Staritz Cornelia. 2010. *Global Value Chains in a Postcrisis World. A Development Perspective*. Washington, DC: World Bank.

Collier, Paul. 2007. *The Bottom Billion. Why the Poorest Countries are Failing and What Can Be Done About It*. Oxford; New York: Oxford University Press.

Compagnie des Bauxites de Guinée. 1983. *Working Together in Guinea*. Pittsburgh: CBG; OFAB.

Deripaska, Oleg. 2014. Quand la discipline s'impose à la métallurgie. La voix de Rusal 76, p. 3.

Dicken, Peter, Kelly Philip F., Olds Kris, and Wai-Chung Yeung Henry. 2001. Chains and Networks, Territories and Scales: Towards a Relational Framework for Analysing the Global Economy. *Global Networks* 1 (2): 89–112.

Dörre, Klaus. 2010. Social Classes in the Process of Capitalist Landnahme. *Socialist Studies/Études socialistes* 6 (2): 43–74.

EITI. 2015. Guinea, http://eiti.org/Guinea/.

European Commission. 2013. New disclosure requirements for the extractive industry and loggers of primary forests in the Accounting (and Transparency) Directives (Country by Country Reporting)—Frequently asked questions. COMM/PRESS/01, http://europa.eu/rapid/press-release_MEMO-13-541_en.htm, updated on 6 Dec 2013, checked on 6 Oct 2013.

European Commission. 2014. Proposal for a regulation of the European Parliament and of the Council setting up a Union system for supply chain due diligence self-certification of responsible importers of tin, tantalum and tungsten, their ores, and gold originating in conflict affected and high-risk areas. Brussels, http://eur-lex.europa.eu/resource.html?uri=cellar:5de359c4-a5f8-11e3-8438-01aa75ed71a1.0002.01/DOC_1&format=PDF, updated on 20 Mar 2014, checked on 28 Jan 2015.

Fayol, Henri. 1929. *Allgemeine und industrielle Verwaltung*. München: Oldenbourg.

Fischer, Karin, Christian Reiner, and Cornelia Staritz. 2010. Einleitung. Globale Güterketten, weltweite Arbeitsteilung und ungleiche Entwicklung. In *Global Güterketten - Weltweite Arbeitsteilung und ungleiche Entwicklung*, ed. Karin Fischer, Christian Reiner, and Cornelia Staritz, 7–24. Wien: Promedia.

Forster, Meinhard. 1976. *Struktur und Risiken der deutschen Nichteisen-Metallversorgung. dargestellt am Beispiel Kupfer, Aluminium, Nickel, Zink, Blei und Zinn*. Hamburg: Weltarchiv.

Gereffi, Gary. 1997. Global Production Systems and Third World Development. In *Global Change, Regional Response: The New International Context of Development*, ed. Barbara Stallings, 100–143. Cambridge: Cambridge University Press.

Global Policy Forum Europe. 2011. Anforderungen an eine zukunftsfähige Rohstoffstrategie. Stellungnahme zivilgesellschaftlicher Organisationen zur Rohstoffstrategie der Bundesregierung. Bonn, http://www.bicc.de/fileadmin/Dateien/pdf/press/2010/Anforderungen_an_eine_zukunftsfaehige_Rohstoffstrategie.pdf.

Habig, Gutmann. 1983. *Möglichkeiten und Grenzen einer Kontrolle internationaler Rohstoffmärkte durch Entwicklungsländer*. Weltarchiv: Das Beispiel des Kupfer- und Aluminiummarktes. Hamburg.

Harvey, David. 2006. *The Limits to Capital*. London, New York: Verso.

Heß, Ralf. 2012. Die deutsche Strategie zur Rohstoffsicherung nimmt Gestalt an. In Telepolis, 29 Mar 2012, http://www.heise.de/tp/artikel/36/36673/1.html, checked on 5 Aug 2014.

Hibou, Béatrice. 1999. La décharge, nouvel interventionnisme? *Politique Africaine* 73: 6–15.

Holloway, Steven Kendall. 1988. *The Aluminium Multinationals and the Bauxite Cartel*. New York: St. Martin's Press.

Hönke, Jana. 2009. Transnational pockets of territoriality. Governing the security of extraction in Katanga (DRC). Working Paper Series of the Graduate Centre Humanities No. 2. Leipzig: Leipziger Universitätsverlag.

Hopkins, Terence K., and Wallerstein Immanuel. 1986. Commodity Chains in the World-Economy Prior to 1800. *Review (Fernand Braudel Center)* X (1): 157–170.

International Finance Corporation. 2012. IFC Performance Standards on Environmental and Social Sustainability, ed. World Bank Group, http://www.

ifc.org/wps/wcm/connect/c8f524004a73daeca09afdf998895a12/IFC_Performance_Standards.pdf?MOD=AJPERES, checked on 20 Jan 2015.

International Monetary Fund. 2008. Guinea: Selected Issues and Statistical Appendix. IMF Country Report 08/20, Washington, DC, checked on 5 Sept 2014.

Jaeger, Nicola. 2012. *Rohstoffpolitik vom Kopf auf die Füsse stellen.* Oxfam Deutschland/Weed: Alternativen zur rücksichtslosen Ressourcenausbeutung im globalen Süden. Berlin.

Jäger, Johannes, and Bernhard Leubolt. 2011. Reprimarisierung als neue Entwicklungsstrategie in Lateinamerika. *Kurswechsel* 3: 59–69.

Jäger, Johannes, Bernhard Leubolt, and Lukas Schmidt. 2014. Alles Extraktivismus in Südamerika? Rohstoffrenten und Politik in Brasilien, Chile und Venezuela. *Journal für Entwicklungspolitik* 30 (3): 9–27.

Jessop, Bob, Neil Brenner, and Martin Jones. 2008. Theorizing Sociospatial Relations. *Environment and Planning D: Society and Space* 26 (3): 389–401.

Kannankulam, John. 2006. *Autoritärer Etatismus im Neoliberalismus.* Hamburg: Zur Staatstheorie von Nicos Poulantzas. VSA.

Kauder, Volker, Hasselfeldt Gerda, and Brüderle Rainer. 2011. Antrag der Fraktionen der CDU/CSU und FDP: Wirtschafts- und Außenpolitik für eine sichere Rohstoffversorgung - Wachstum und Arbeitsplätze in Deutschland Europa und den Partnerländern. Drucksache17/7353. Berlin: Deutscher Bundestag.

Kerkow, Uwe, Martens Jens, and Müller Axel. 2012. Vom Erz zum Auto. Abbaubedingungen und Lieferketten im Rohstoffsektor und die Verantwortung der deutschen Automobilindustrie. Aachen, Bonn, Stuttgart: MISEROR, "Brot für die Welt", Global Policy Forum Europe.

Knierzinger, Johannes. 2011. Chieftaincy and Development in Ghana: From Political Intermediaries to Neotraditional Development Brokers. Working Papers of the Department of Anthropology and African Studies of the Johannes Gutenberg University Mainz 124, Mainz.

Knierzinger, Johannes. 2014a. The Socio-Political Implications of Bauxite Mining in Guinea: A Commodity Chain Perspective. *The Extractive Industries and Society* 1: 20–27.

Knierzinger, Johannes. 2014b. Überlegungen zu Steuerung und ungleichem Tausch in Rohstoffgüterketten mit Beispielen aus Guineas Bauxitbergbau. *Journal für Entwicklungspolitik* 30 (3): 76.

Krause, Alexandra. 2002. Die EU als internationaler Akteur in Afrika. *Aus Politik und Zeitgeschichte B* 13–14/2002: 24–30, Checked on 20 Jan 2017, http://www.bpb.de/system/files/pdf/NX4E6Z.pdf.

Küblböck, Karin. 2014. Can 'Undistorted Access' Lead to Inclusive Development? The EU Raw Materials Initiative and Possible Effects upon

Resource-Based Development in Africa. *Journal für Entwicklungspolitik* 30 (3): 89–106. [ISBN 978-3-902996-02-2; ISSN 0258-2384].
Küblböck, Karin, and Pinter Silke. 2015. *Gläserner Bergbau? Transparenzinitiativen im extraktiven Sektor. Briefing Paper 11.* Wien: ÖFSE.
Lambert, Tobias. 2012. Ressourcen für Europa. Die Rohstoffstrategien der EU und Deutschlands gehen auf Kosten des globalen Südens. In *Der Neue Extraktivismus. Eine Debatte über die Grenzen des Rohstoffmodells in Lateinamerika*, ed. FDCL, Rosa-Luxemburg-Stiftung, 128–144. Berlin: FDCL-Verlag.
Larrue, Jacques. 1997. *Fria en Guinée. Des aspects humains d'une industrialisation differente.* Paris: Karthala.
Maier, Charles S. 2006. Transformations of Territoriality 1600–2000. Space, Place, Territory. In *Transnationale Geschichte. Themen, Tendenzen und Theorien*, ed. Gunilla-Friederike Budde, 32–55. Göttingen: Vandenhoeck und Ruprecht.
Mann, Michael. 1984. The Autonomous Power of the State: Its Origins, Mechanisms, and Results. *European Journal of Sociology* 25 (2): 185–213.
Mayes, Robyn. 2015. A Social Licence to Operate: Corporate Social Responsibility, Local Communities and the Constitution of Global Production Networks. *Global Networks* 15 (1): 109–128.
Mishel, Lawrence, and Davis, Alyssa. (2014). CEO Pay Continues to Rise as Typical Workers are Paid Less. *Issue Brief #380.* Washington, D.C.: Economic Policy Institute.
Moran, Theodore. 2013. A Little More Transparency at EITI Can Go a Long Way. In Center For Global Development, 24 May 2013, http://international.cgdev.org/blog/little-more-transparency-eiti-can-go-long-way, checked on 6 Oct 2013.
Natural Resource Governance Institute. 2013. Financials, http://www.revenuewatch.org/about/financials, updated on 6 Oct 2013, checked on 6 Oct 2013.
Nodalis; Sedec; mtbs. 2013. *Développement des infrastructures auxiliaires de transport minier en Guinée Conakry. Rapport de diagnostic - previsoire.* Conakry: ANAIM.
Norton Rose Fulbright. 2013. Amendments to the Guinean Mining Code, http://www.nortonrosefulbright.com/knowledge/publications/99188/amendments-to-the-guinean-mining-code, checked on 20 Jan 2015.
Open Society Institute of Southern Africa et al. 2009. Breaking the Curse: How Transparent Taxation and Fair Taxes Can Turn Africa's Mineral Wealth into Development. Johannesburg, http://www.christianaid.org.uk/Images/breaking-the-curse.pdf, checked on 27 Jan 2015.
Oumar, Alpha. 2014. Mines: Convention AMC, le député Dr Fodé Maréga exige des explications. In aminata.com, 8 July 2014, http://aminata.com/

mines-convention-amc-le-depute-dr-fode-marega-exige-des-explications/, checked on 14 Jan 2015.

Pezet, Anne, Loison Marie-Claire, and Berrier Céline. 2009. Corporate Social Responsibility before CSR. Practices at the Company Aluminium du Cameroun, 1950s–1970s.

Pimmer, Stefan, and Lukas Schmidt. 2015. Dependenztheorien Reloaded (Einleitung). *Journal für Entwicklungspolitik* 31 (3): 4–10.

Plank, Leonhard, and Cornelia Staritz. 2010. Globale Produktionsnetzwerke und "prekäres Upgrading" in der Elektronikindustrie in Mittel- und Osteuropa. In *Global Güterketten - Weltweite Arbeitsteilung und ungleiche Entwicklung*, ed. Karin Fischer, Christian Reiner, and Cornelia Staritz, 179–201. Wien: Promedia.

PWYP. 2015. Practice What You Preach: Financial Transparency, http://www.publishwhatyoupay.org/about/funding, checked on 27 Jan 2015.

Radhuber, Isabella M. 2015. Extractive Processes, Global Production Networks and Inequalities. DesiguALdades.net Working Paper Series 89. Berlin: International Research Network on Interdependent Inequalities in Latin America.

Raikes, Philip, Jensen Michael Friis, and Ponte Stefano. 2000. Global Commodity Chain Analysis and the French Filière Approach: Comparison and Critique. Copenhagen: CDR Working Paper.

République de Guinée. 2011. Code Minier de la République de Guinée. Conakry, http://itie-guinee.org/codes_miniers/Code_Minier_2011.pdf, updated on 25 Oct 2011, checked on 16 May 2014.

République de Guinée. 2013. Code Minier 2011 amendé. Conakry, updated on 25 Oct 2011, checked on 16 May 2014.

Revenue Watch Institute. 2013. Guinea Adopts Amended Mining Code After Engaging With Stakeholders, 12 May 2013, http://www.revenuewatch.org/news/blog/guinea-adopts-amended-mining-code-after-engaging-stakeholders, checked on 4 Oct 2013.

Revenue Watch Institute. 2014. La chaîne de valeur dans les industries extractives. Atelier de formation des formateurs sur le suivi des obligations légales des sociétés minières. Kindia. Powerpoint presentation.

Revenue Watch Resource Center. 2015. The Value Chain, http://archive-2011.resources.revenuewatch.org/en/backgrounder/value-chain, checked on 22 Jan 2015.

Robertson, David. 2007. Mining giants indignant that someone else should rape and pillage Africa. In *The Times*, 29 Jan 2007, http://www.gata.org/node/4757, checked on 30 Mar 2015.

Ross, Michael L. 1999. The Political Economy of the Resource Curse. *World Politics* 51: 297–322.

Rubbers, Benjamin. 2010. Claiming Workers' Rights in the Democratic Republic of Congo: The Case of the Collectif des ex-agents de la Gécamines. *Review of African Political Economy* 37: 329–344.

Sachs, Jeffrey, and Warner Andrew M. 1995. Natural Resource Abundance and Economic Growth. Cambridge, MA.: National Bureau of Economic Research (NBER working paper series).

Shafer, D. Michael. 1986. Undermined: The Implications of Mineral Export Dependence for State Formation in Africa. *Third World Quarterly* 8 (3): 916–952.

Smet, Koen, and Markus Seiwald. 2014. Bergbau als Entwicklungschance? Überlegungen anhand der Beispiele Ecuadors und Südafrikas. *GW-Unterricht* 136 (4): 5–14.

Soumah, Ibrahima. 2008. *The Future of Mining Industry in Guinea*. Paris: L'Harmattan.

Southall, Roger, and Henning Melber (eds.). 2009. *A New Scramble for Africa? Imperialism, Investment and Development.* Scottsville, Kwazulu-Natal: University of KwaZulu-Natal Press.

Strange, Susan. 1996. *The Retreat of the State. The Diffusion of Power in the World Economy.* Cambridge: Cambridge University Press.

Taylor, Magnus. 2011. Paul Collier: Can Africa harness its resources for development? In African Arguments, 1 Nov 2011, http://africanarguments.org/2011/11/01/paul-collier-can-africa-harness-its-resources-for-development/, checked on 25 July 2013.

Taylor, Peter J. 2000. World Cities and Territorial States Under Conditions of Contemporary Globalization. The 1999 Annual Political Geography Lecture. *Political Geography* 19: 5–32.

UNCTAD. 2007. World Investment Report 2007. Transnational Corporations, Extractive Industries and Development. New York, Geneva: United Nations, http://unctad.org/en/docs/wir2007_en.pdf, checked on 28 Aug 2014.

Wann, Ousmane. 2014. *Guide de suivi des obligations légales des sociétés minières.* Conakry: Workshop reader.

Williams, Martin. 2010. The Gold Standard of Governance: Mining, Decentralization, and State Power in Senegal. *Politique Africaine* 117: 127–149.

WTO. 2013. Dispute Settlement: Dispute DS394 China–Measures Related to the Exportation of Various Raw Materials, http://www.wto.org/english/tratop_e/dispu_e/cases_e/ds394_e.htm, checked on 16 Sept 2013.

CHAPTER 3

Case Study: Guinea

Since the late 1950s, the needs of the global production network of aluminum have resulted in extensive social transformations in Africa. In Guinea, the most visible outcomes of this process are the mining towns of Fria, Sangaredi, Kamsar, and Débélé. In addition, the Loose Islands near the capital Conakry were for a long time (in operation until 1965) the only Guinean source of bauxite and thereby also significantly contributed to the growth of Conakry. All of these agglomerations have been planned, created, and maintained by transnational corporations and have seen the most rapid population growth in Guinean history. Similar developments can be observed in Sierra Leone and Ghana, but in these cases, the mutual dependence of firms and raw material providing countries (and thus the consequences for the population) was less important than in Guinea (Fig. 3.1).

Guinea became independent in 1958, after a referendum against General de Gaulle's Communauté française, a reformed continuation of the French colonial empire. When de Gaulle visited Guinea a few weeks before the referendum, Guinea's first revolutionary leader Sékou Touré made a fiery speech that also contained his famous statement that his country preferred "poverty in liberty to riches in slavery". De Gaulle was booed by the Guinean crowd and responded—apparently in anger—that Guinea was free to opt out, and guaranteed to accept the decision of the population. On 2 October 1958, only 4 days after the referendum, Guinea was formally granted independence. The media thereafter depicted a personally offended de Gaulle, who withdrew all French

J. Knierzinger, *Bauxite Mining in Africa*, International Political Economy Series, DOI 10.1007/978-3-319-52706-2_3

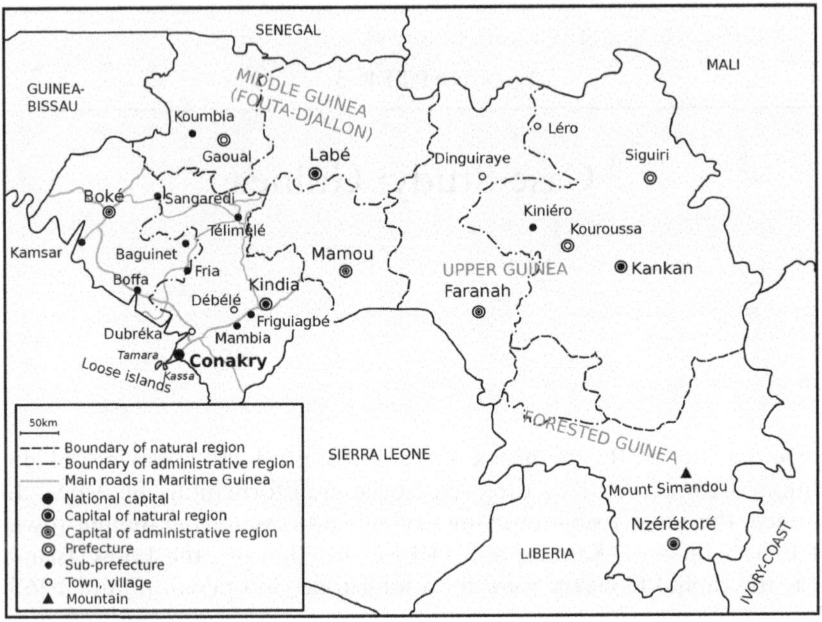

Fig. 3.1 Map of Guinea

personnel immediately and prevented Guinean ships with cash crops from entering French harbors. He also tried to obstruct the admission of Guinea to the United Nations (cf. Balde 2005: 9). From 1965 to 1975, France had no ambassador in Guinea and was officially represented by Italy. As the regime became increasingly oppressive, the Italian embassy also became less active and France increasingly used the managers of Pechiney as intermediaries in its negotiations with the Guinean government (Larrue 1997: 196; cf. Uwechue 1991: 755).

De Gaulle's ostentatious anger over Guinea's opt-out was understandable. Due to the uprisings in North Africa and Indochina, France had undergone a serious political crisis during the Fourth Republic from 1946 to 1958. During this period, French governments only remained in power about 6 months on average (cf. Grilli 1994: 6–8; Martin 1982; Schmidt 2007). Guinea was a showcase for achieving the definitive end of the French empire in Sub-Saharan Africa, and even rejected the dreams of prolonging this adventure under the new territorial configuration of the European Union. The country thereby played an important role in the non-aligned movement (e.g., through its role in the formation of the International Bauxite Association (IBA), see Holloway 1988: 52–54). However, cutoff from trade in cash crops that had filled the state coffers and the pockets of thousands of petty farmers until the retreat of the colonialists, the revenues of the new socialist government now almost exclusively came from three bauxite mines controlled by a global aluminum oligopoly (cf. Campbell 2009a).

Guinea's first important mining complex—Fria—started to produce shortly after the independence and was owned (with changing shares) by the Guinean government, Pechiney (France), the Olin Mathieson Chemical Corporation (later Noranda, USA), Alusuisse (Switzerland), British Aluminum, VAW (Germany), and Alcan (Canada). The Compagnie des Bauxites de Kindia in Débélé started production in 1974 and was constructed with the help of Russian experts. It exported almost exclusively to the Soviet Union and was fully owned by the Guinean state. The largest African producer, the Compagnie des Bauxites de Guinée, comprised the cities of Sangaredi and Kamsar and started to export bauxite in 1973. It belonged, also with changing shares, to the Republic of Guinea, Alcan (Canada), Alcoa, Harvey, Martin Marietta, Reynolds (all US), Pechiney (France), VAW (Germany), and Montecatini Edison (Italy). This extensive list of companies contains five of the so-called Six Sisters—Alcoa, Reynolds, Kaiser (all of them based on the US), Alcan (Canada), Alusuisse (Switzerland), and Pechiney (France)—producing about 60% of worldwide primary aluminum at that time (USGS 2016). All the firms and state enterprises working together in Guinea produced almost 70% of worldwide primary aluminum in 1960, about 60% in 1975 (cf. USGS 2016), and again more than 75% in 2014 (if we count China's production as one state enterprise).

The needs of this large industrial network resulted in the largest demographical restructuration of Guinean history: Fria-*Centre* grew from 5000 inhabitants in the 1960s to about 90,000 in 2012 (Interview with Hospital director 1/04/2014);[1] Kamsar from a fishing village with several hundred inhabitants to 8000 in 1973 and 113,000 today (République de Guinée 2014: 7);[2] and Sangaredi was apparently almost entirely deserted until the 1960s and had grown to 70–80,000 in 2014. The aluminum town of Edéa in Cameroon (Alucam and Socatral), which received alumina from Fria, grew from about 8000 people in the 1950s to about 240,000 today (Encyclopedia Britannica 2013; Pauthier 2002:

27; Pechiney Ugine Kuhlmann 1982: 4–5). This demographic aspect of bauxite mining and processing also corresponds to the companies' massive need for energy. Today, Alucam still accounts for more than 34% of total power consumption in Cameroon (Husband et al. 2009: 30), and the industrial complex of Fria consumed as much petrol as the rest of Guinea until its shutdown in 2012 (World Investment News 2009). In the 1970s, aluminum smelting in Ghana accounted for some 60% of the country's electricity consumption (Riverton 2014: 36).

The political implications of this new economic direction can be compared with many other African countries being dependent on the extractive industry (Agnew and Grant 1996: 738): depending almost exclusively on several thousand open pit miners, the political elite became increasingly alienated from its population and reinforced the country's extractivist path—at the expense of agriculture, processing, and services. The neoliberal turnaround in 1984, when Sékou Touré passed away and Lansana Conté took over power, only catalyzed these tendencies in Guinea.

Conté had already been one of the most influential militaries under Touré and came to power in a bloodless coup shortly after Touré's death, as the president of the Comité Militaire de Redressement National (CMRN). Conté, suspended the country's constitution and immediately resumed negotiations on structural adjustment with the IMF and the World Bank, which had already been started in 1982 (Balde 2005: 18; Campbell and Clapp 1995: 431).

Naturally, the political turnaround in Guinea was strongly welcomed by the aluminum companies. Several other new ministers had been working for the World Bank before (Larrue 1997: 282). Real tax revenues started a steady decline until the 2000s (cf. International Monetary Fund 2008: 9) and are today most likely still lower than in the early 1980s. Out of 131 existing public enterprises, 25 were privatized and 68 were closed (Balde 2005: 24). The new austerity measures led to continuous violent protests since 1986. Alpha Condé, the current president of Guinea—who had already been a leftist critic of the Touré regime—agitated from abroad against the Conté regime at this time and was presented as the new resistance leader by the French press (Condé 1972; cf. Larrue 1997: 155, 208). Conté reacted to these protests by firing ministers and reducing the income of government officials, through monetary assistance for transport and food, as well as increasing the salaries of the remaining civil servants (Balde 2005: 26; Dux 2000). In 1991, being confronted with continuing strikes and protests, Conté raised public sector salaries by 100% and thereby prompted the IMF to suspend negotiations for several months. In this year, Guinea ranked last in the human development index (Campbell and Clapp 1995: 447). In 1996, a military revolt even destroyed the presidential offices and killed dozens of Guineans. Conté was taken to the main military camp of Conakry and was again forced to promise reforms, this time also regarding the military forces. Instead of keeping these extorted promises, Conté arrested hundreds after his release.

In 1995, the government also published a new mining code that radically reduced taxes. Its general impact was negligible because it was neither respected during the formulation of new contracts nor did it change the contracts of the existing projects (Diallo et al. 2011: 48; International Monetary Fund 2008: 9). The only success that could be put in relation with the code is the start of industrial gold mining (cf. Windfuhr 2004: 16). In recent years, one of these new gold mines has started to significantly contribute to government revenue. However, before this start of several new gold mining projects, taxes from mining considerably declined, in both absolute and relative terms. While mining was responsible for 79% of tax revenues and 95% of exports in 1984 (World Bank 1990: 32), tax revenues from mining decreased to 40% in 1993 (Campbell and Clapp 1995: 427) and 20.5% from 2002 to 2005 (International Monetary Fund 2008; cf. Dux 2000). Campbell mainly relates the absolute decline of mining tax revenues to new tax arrangements between CBG and the government of Guinea, which entered into force in 1991 and influenced similar arrangements in Fria. In both the cases, a bauxite levy was replaced by taxes that were tied to the—falling—aluminum prices and additionally contained clauses that would have led to a further decline if aluminum prices would have risen again (Campbell and Clapp 1995: 431, 449). At the same time, the government succeeded in boosting revenues from other sources, such as wages, profits, goods and services, and international trade (Campbell and Clapp 1995: 436). However, as bauxite and alumina alone still accounted for about 71.6% of exports from 1990 to 1993 and 61.5% from 2002 to 2005 (International Monetary Fund 2008), a good part of these new taxes have to be understood as shifts of taxation within the same sector from taxes on capital to taxes on labor. Guinea neither significantly diversified its exports during this time (the only exception is again gold) nor did it achieve industrial growth. The SAPs rather led to the reversal of an industrialization trend that has been discernible since the 1960s (cf. Dux 2000). After another government shuffle in 1998, the regime Conté reached a stage of profound corruption and cronyism, associated with accelerated economic decay. Drug and mafia networks crept into public institutions (cf. Trade unionist 17/04/2014; Larrue 1997: 307). An activist reported that neither the Supreme Court nor the administration worked anymore:

"You had no institution anymore in this country that was still functioning. And nobody believed anymore in these institutions" (Interview with Activist 2/03/2014; translation by JK; cf. Christian leader 14/02/2014).

As a result, the IMF, the World Bank, the African Development Bank, and the European Union stopped economic assistance to Guinea in 2002 and 2003 and thereby suspended the HIPC Initiative (Committee on foreign affairs 2007: 52; Interview with Activist 2/03/2014; Legal expert 13/03/2014). Together with the new mining boom that started in the mid-2000s, this fragile political situation led to a series of resistance movements.

In 2007, after about two decades of nepotism and economic decline, the strong Guinean trade unions and other civil society activists pushed for political change which eventually led to the coming into power of a military junta, the National Council for Democracy and Development (CNDD), from 2008 to 2009. The CNDD was initially supported by the majority of the social movements and was led by Moussa Dadis Camara, the previously little known Chief of Fuels at the Guinean army base in Kindia. (Arieff and Cook 2010). Dadis Camara's inconsistent, brutal, but nevertheless (at least temporarily) quite popular rule soon led to international isolation. On 28 September 2009, Guinean security forces opened fire on civilian demonstrators in Conakry who were protesting against the CNDD and Dadis Camara's intentions to run for president, killing over 150, injuring many more and raping numerous women (Arieff and Cook 2009). On 3 December 2009, Dadis Camara was shot and seriously wounded by a member of his own presidential guard. He was evacuated to Morocco for medical treatment. In Dadis Camara's absence, the CNDD defense minister, Sékouba Konaté, informally assumed the position of acting head of state (Arieff and McGovern 2013).

Only after another 12 months of transition, the country received probably its first democratically elected president, the former opposition leader, and long-time expatriate Alpha Condé. However, both the proudly announced economic take-off and the political renaissance lost momentum due to the global economic downturn that followed the subprime crisis of 2007/2008. In 2013, only several months after implementing a fairly progressive mining code which imposed higher taxes and local development efforts, Alpha Condé felt the need to revise this code again, thereby giving into the pressure of mining majors and the World Bank (cf. Knierzinger 2014: 25).

At present, only five industrial mines—two bauxite and three gold mines—are operational in Guinea. These mines belong to CBG, CBK, Semafo (a Canadian company that started to produce in Kiniéro in 2002), SMD (Société Minière de Dinguiraye) in Léro, 85% owned by the Norwegian Guinor Gold Corporation, 15% held by the state, and SAG (Société Ashanti de Guinée) in Siguiri, acquired by Ashanti Goldfields in 1998, which became Anglogold Ashanti in 2004. Semafo, SMD, and SAG are goldmines. In 1984, the *Association pour la recherche et l'exploitation du diamant et de l'or* (Aredor) began to mine diamonds in Banankoro, with changing shares held by the Guinean state, Swiss, Australian, British, Canadian, Guinean, and Pakistani investors. The mine has been abandoned several times, for the last time in 2008.[3]

All of these mines have been created before the mining boom of the 2000s. Since then, in spite of numerous promises, no new bauxite or alumina projects have started to produce. If the announcements in the book of the former mining minister Ibrahima Soumah in 2008 had come true, three new alumina plants would have started production already in 2008 and 2010 (Soumah 2008; cf. Guinéenews 2015). According to the current Guinean government, bauxite production was to increase from around 17 million tons in 2013 to 61 million in 2016/2017, and within the same time span, alumina production was to rise from zero to 16–20 million tons. The same publication counts eight new bauxite mining projects that are currently underway (Bureau de Presse de la Présidence 2013; cf. The Express Tribune 2012; World Investment News 2009). According to a World Bank study for the Agence Nationale d'Amenagement des Infrastructures (ANAIM), if all projects developed as planned until 2020, Guinea would produce more than 100 million tons of bauxite, or 40% of 2011 worldwide bauxite production (Nodalis et al. 2013: 7, 25). In 2007, the government also announced that it was no longer giving out permits for sole bauxite mining (Banque mondiale 2007). However, among the numerous projects that have been allowed since then, only one—a co-project of Alcoa and Rio Tinto in Kamsar—seems to have the potential to lead to a new alumina refinery. All the other announcements have been recalled or delayed, recently also due to the Ebola outbreak.

In the following, the two currently existing spheres of influence of transnational aluminum companies in Guinea will be analyzed: on the one hand, the Russian sphere comprising Fria and Débélé and on the other hand, the US/Australian influence through Alcoa and Rio Tinto that extends over Sangaredi and Kamsar. These two spheres will be treated separately in a chronological order and subsequently compared in terms of their entanglements with the formal Guinean political system. The available sources on these two corporate spaces are quite different. In the case of Fria, some historical literature exists and I was additionally able to draw upon the unpublished writings of Ibrahima Diallo, a Guinean intellectual who lives in Fria. In the case of CBG, I more strongly focused on the structural analysis of the current situation.

3.1 The Decision Space of Rusal in Guinea

Rusal was formed in 2000 through a fusion of Sibirsky Aluminium and Sibneft after a chaotic decade of fights—called the aluminum wars—over the aluminum stakes of the former Soviet Union. This "restructuring" of the industry claimed the lives of over one hundred plant managers, metal traders, and journalists. The ultimate winners of these fights were the oligarch Roman Abramovich (today owner of the Chelsea Football Club) and Oleg Deripaska, an intimate of Vladimir Putin and married to a granddaughter of Boris Yeltsin (cf. Fortescue 2006: 80). Strongly supported by the Russian state, Deripaska became the majority stakeholder of Rusal and thereby the single most influential actor in the global aluminum business.[4] In 2007, Rusal took over the Russian aluminum company Sual and the alumina assets of the mining multinational Glencore to become the world's largest aluminum producer. Sual was owned by another oligarch, Victor Vekselberg, who also ranks among the richest Russians. Vekselberg left Rusal in March 2012—1 month before Friguia closed its doors—because he did not succeed in convincing Deripaska to refinance the heavily indebted company by selling Norilsk Nickel, a nickel and palladium site in Russia. Deripaska had bought Norilsk shortly before the sub-prime crisis and is reported to have kept it due to political ambitions, despite Rusal's debts of 11 billion USD in 2012 (Neue

Zürcher Zeitung 2010; Khrennikov 2012; Revoin 2012). Deripaska himself was supposed to be the richest Russian before the sub-prime crisis and lost more than 20 billion Euros in the following years, more than any other Russian oligarch during the crisis (Le Figaro 2008; Harding and Macalister 2010). A Guinean newspaper reports on debts of 16.8 billion USD that have been rescheduled until 2020 in accordance with 70 creditors (Diallo 2014f). If Rusal had sold Norilsk, it could have evaded a good part of these procedures. Besides, this could have fundamentally changed the fate of the population of Fria.Rusal currently counts 69,000 employees and disposes over 15 aluminum smelters, 11 alumina refineries, eight bauxite mines, three aluminum powder plants, two secondary aluminum plants, and four foil mills. The bauxite mines are situated in Guinea, Jamaica, Guyana, and Russia. According to Rusal's website, the *Compagnie des Bauxites de Kindia* provides 30% of Rusal's bauxite (Rusal 2014b). Besides this core sector, Deripaska owns many other factories—including Russia's largest car manufacturer, GAZ with 18 manufacturing facilities—and has worldwide shares in the mining sector and other sectors, such as aviation, construction, mechanical engineering, electricity generation, banking, and insurance (Bornefeld 2007; Brüggmann 2007; Die Presse 2014; Der Standard 2014; Khrennikov 2012).

The two sites in Guinea are connected to a refinery in Nikolaiev, Ukraine, the Alscon smelter in Nigeria (which has only been operable for several years; see Sect. 1.2), and a refinery in Aughinish, Ireland, which has initially received Guinean bauxite from Kamsar through the production network of CBG. This site in Aughinish had been bought by Glencore in 1999, which was in turn bought by Rusal in 2007 (Bouliguine 2007; Committee on foreign affairs 2007: 41; Rusal 2014b). About one-third of the Irish alumina is delivered to smelters in Germany (Hütz-Adams et al. 2014). In 2015, as a consequence of the Crimea crisis, the refinery in Nikolaiev was nationalized by the Ukrainian government (The Moscow Times 2015). No site of Rusal in Africa is ISO certified.[5]

Besides the facilities in Fria and Débélé, Rusal has also recently acquired mining rights in Dian-Dian, north of Sangaredi.[6] As the reserves in Débélé will be depleted in the coming years, these reserves are a fundamental basis for the whole production chain of Rusal. In addition, they probably also prevented a total debacle after Rusal's initial public offering at the stock exchange of Hong Kong in 2010. As the first non-Asian company, Rusal decided to go to Hong Kong because of its stakes in China, but due to the poor outlook of the aluminum sector in general and Rusal, in particular, its stocks slumped 43% from January 2010 to March 2012.[7] This slump was probably catalyzed by a public announcement of the Guinean mining minister Mamadou Thiam at the Hong Kong stock exchange in January 2010, claiming that Rusal's declared stakes in Guinea were in fact much lower. The regime under Dadis Camara had nullified the privatization of Fria of 2006 and claimed 860 million USD from Rusal in compensation for the losses since then. Thiam told the Financial Times that "[s]ubscribers should know that, barring a settlement with Guinea, the value of the Rusal stock they acquire might very well be overstated" (Burgis et al. 2010). From Rusal's viewpoint, the decision of the Guinean government to introduce a new mining code in 2011 came exactly at the wrong moment. The heavily indebted company already struggled with similar mutinies in Nigeria and Guyana (BBC News 2012; Diallo 2013l; Marshall 2013) and still faced a fall of the aluminum price, which coincided with rising energy prices (Deguenon and Traoré 2011d; Diallo 2014g).

3.1.1 Fria

The area of Fria was scarcely populated in the 1950s when the colonial power created a new prefecture[8] and began to build a town for 20–30,000 inhabitants (Diallo 2010: 6; Former executive Rusal 3/04/2014). Already in the mid-1960s, an architectural bureau had to be hired to reduce the rapidly growing slums (about one-third of the population) around the *cité* of Fria (Pauthier 2002: 16–23, 90–97). From its inauguration in 1960 until 2012, Fria remained the only alumina refinery on African soil (Bah 2006). In 2014, the urban municipality (*commune urbaine*) of Fria-*Centre* counted 61,691 inhabitants (République de Guinée 2014: 7),[9] all of whom are directly or indirectly dependent on the mining complex and were hit hard by the halt of production since April 2012 that has put the town—according to a spokeswoman of Rusal—"on the brink of a humanitarian catastrophe" (Radio Netherlands Worldwide 2012). With the exception of the market area, two hospitals, and three story buildings that need electricity to pump up the tap water, the city has been deprived of power since 18 January 2013 (Interview with Activist 3/04/2014; Journalist 15/01/2014; Mayor of Fria 18/01/2014; Diallo 2013b).[10] In addition, Rusal has considerably reduced the quality of health care and other social services. Due to polygamy, the workers' families often comprise several households and are additionally charged with children from relatives. With the abrupt cut of the workers' salaries, these families ran into serious troubles, being forced to take kids out of school or send them to relatives in other cities. Rise consumption per person declined by about one-third (see below). The mayor of the town, Amara Traoré, stated that Fria "is in danger of disappearing if the factory stops. [...] Today I've got a starving population. People are selling their property, their homes, plots of land and even furniture to survive" (Radio Netherlands Worldwide 2012). Since the takeover of Rusal at the turn of the millennium, most of Fria's output went to Russia, from 2007 until the stop of production in 2012 mostly to the Khakas Aluminium Smelter in Khakassia (Interview with Former executive Rusal 4/04/2014; Diallo 2010: 36) and from there mostly to the Russian car industry (Interview with Journalist 15/01/2014). The change of ownership during the 1990s thereby led to an extensive reshuffle of the supply chains and also necessitated technical adaptions. As in the case of alumina and bauxite, the aluminum smelters are technically geared to the kind of alumina that they receive (Interview with Former executive Rusal 25/03/2014).

3.1.1.1 Fria Under Pechiney

During WWII, the French government sought to replace shrinking imports from the US by proper aluminum capacities. This was done partly in France and by constructing a full chain of production between Fria (bauxite and alumina) and Edéa (aluminum) in Cameroon. Prospection for this project already began in the 1940s (cf. Pauthier 2002: 27) and the project development started officially in July 1951 with the foundation of the *Société africaine de recherche et d'études pour l'aluminium* (SAREPA) based on Conakry, which initially had the aim of installing a fully integrated aluminum industry in Guinea (Ribadeau-Dumas 1952: 17). In the 1970s, Fria thereby connected three incomplete African production chains with one another: Ghana received alumina from Guinea despite proper bauxite mines; Cameroon had bauxite reserves, but no mines and received most of its alumina from Guinea until the takeover of Rusal in Fria; and Guinea has an alumina refinery but no smelters (Senghaas-Knobloch 1975: 286; see also Habig 1983: 154). Besides Cameroon and Ghana, the alumina produced in Fria was also shipped to Spain,[11] Nigeria,[12] Brazil, and Russia.[13] In

1958, Pechiney additionally began to construct another aluminum site and a huge dam (1800 km^2)[14] on the river Kouilou in French Congo[15] together with Alcoa that should also have received bauxite from Fria (Vitry 1957). However, with the coming into power of the revolutionary Alphonse Massamba-Débat in 1963, the project was stopped and has never been taken up again (see Tourot 2003).

From the standpoint of the French government, Fria was to become the economic capital of the country and thus a showcase for planned industrialization in Africa. The spirit of the time was both missionary (in a religious and a technocratic sense) and desperate, as the French empire was about to lose its grip in both Africa and Asia. The self-portrait of Pechiney fitted very well into this picture, as it emphasized social and community building policies. The anthropologist Céline Pauthier explicitly mentions the encyclical *Rerum Novarum* as the basis for this orientation. This influential encyclical had been issued by the "worker's Pope" Leo XIII in 1891. Leo XIII underlined the necessity of organic and harmonious relations between workers and employers and was critical of governmental intervention in economic affairs. To the post-war leaders of Pechiney, social progress seemed to be mainly a combination of economic growth and moral enhancement. Pechiney wrote in a memorandum in 1957 that the "moral value of this scheme" is that its African employees would become "agents in spreading human and social development" as "distributors of purchasing power [...] to the benefit of local and regional craftsmen and food crop growers" (Pezet et al. 2009: 27).[16] The Alucam Bulletin—destined for the workers of the plant in Edéa—published articles on topics, such as hygiene, diet, education, work ethics, and polygamy, even featuring a "good housewife" section, where social workers advised the wives of workers. Particularly, the household was seen as a means for the "promotion sociale" of the masses with the clear aim of introducing the French ideal of the core family (see Morgaut 1960: 14).[17] Besides these "moral issues", the Bulletin clearly focused on increasing productivity:

> And you ask yourself: why all that agitation? Why all that noise which seems to be an eternal monotony? Well! Some people have realized that all this movement, all these manifestations, all these delights, constitute a DUTY, or even better, a LAW that obligatorily rules over all living creatures. This duty, this law, which is the real human IDEAL, is called WORK. (Tcheby 1960: 21; translation by JK, accentuation by Tcheby).

According to Pezet et al., the Alucam Bulletin thereby became "a means of social and cultural conditioning" with the aim of adjusting the workers increasingly to "western habits" (Pezet et al. 2009: 29). The workers' journals of Pechiney can thus be read as perfect study examples of the diffusion of Weber's spirit of capitalism, all the more as Pechiney's capitalist mission was at the same time profoundly Christian. This missionary character of Pechiney can still be discerned today. Only in 1997, Jacques Larrue still wrote without irony about the need for a cultural adjustment program and the creation of an "African ant" that could transform the African continent into a "low prices zone" ("en zone de *bas prix*"; italics by Larrue; Larrue 1997: 227).

Already in the late nineteenth century, the predecessor of Pechiney had started to create facilities in deserted areas and was forced to care for the needs of the employees in a very comprehensive way.[18] Pechiney then made similar experiences in projects in the US and Norway in the 1910s, as well as in Italy, Spain, and the Soviet Union during the interwar period (Pezet et al. 2009: 3). This longstanding corporatist tradition of catholic coinage[19] was reinforced by the shift of the late French colonial policies (cf. Pauthier 2002: 4, 16–23). In December 1956, both the company Fria was formed and the *Mission d'aménagement de Guinée* (MARG) began with the construction of the needed infrastructure (Dux 2000; Pauthier 2002: 8). However, the consent of the *Assemblée Territorale* of *Afrique-Occidentale française* (AOF) to this project only came on 17 April 1958, a few months before Guinea gained its independence on 2 October. Despite its leading role, Pechiney only held a share of 26.5% in Fria, which was reduced again in the 1970s and 1980s. Other shareholders were the German VAW (5%), British Aluminium, Alusuisse (each of them holding 10%), and the US firm Olin Mathieson Chemical Corporation with a share of 48.5% (Campbell 1991: 33). The leading role of Pechiney mainly resulted from the fact that the infrastructure of the project was co-financed by France. These infrastructure investments included the extension of the harbor in Conakry, a road from Fria to the existing coastal route, a railway from Fria to Conakry, and the mining town (Pauthier 2002: 1–8, 51–52, 60–61; Habig 1983: 88). The mining complex was to be solely financed by the shareholders, whereas parts of the infrastructure were later also financed by the Guinean government, which in turn heavily relied on foreign donors. After independence, dogged renegotiations about the infrastructural needs of Fria resulted in the government's decision in December 1959 to contribute to publicly usable parts of the corporate infrastructure (Pauthier 2002: 69–75).

In the beginning, Pechiney wanted to found Fria together with Italian partners; however, lacking in financial resources and apparently due to fierce competition with the Canadian firm Alcan, the above list of shareholders became the owners of Fria. Similar to Rusal in the 2000s, Alcan pressured the political authorities by stating that it would be the fault of Pechiney if its Guinean subsidiary—the *Bauxites du Midi*—would not construct an alumina refinery as promised. However, the Guinean government did not (immediately) give in and Alcan was not admitted to Fria. *Bauxites du Midi*, which never exported industrially significant quantities of bauxite, was nationalized later on, because it did not meet the deadline to construct the said refinery (see Larrue 1997: 114). This rivalry can be explained by the geopolitical competition between Europe and the US over the resources in Guinea and the rest of Africa. Despite their private ownership, both companies were highly entangled with their respective governments. The owners of Alcan—the Mellon and Davis families—also possessed the US monopolist Alcoa and were also part of the political establishment in both countries (for Pechiney, see Larrue 1997: 116).

Before the start of production of CBG, Fria was the most important asset of the government, accounting for almost 60% of Guinean exports (Campbell 2009b: 75). According to an audit in the 1980s, Fria produced 7.32 million tons of alumina to the value of about 625 million USD from 1960 to 1973 and paid 46 million USD in taxes during the same period (Diallo 2014f).[20] Due to its *naissance* during a historic turning point, its already particular configuration as a mining town became reinforced by the dwindling public resources. While the private company Pechiney was among the few French operators that did not leave when Touré ostentatiously chased away the colonialists in 1958, the finances of its French infrastructural supporter—the *Mission d'Aménagement Régional de Guinée* (MARG)—faded out within a few years.[21] MARG reported to the *ministère de la France d'outre mer (FOM)*[22] and coordinated both research on developmental issues and the creation of social, administrative, and physical infrastructure in Fria (Pauthier 2002: 56; cf. Larrue 1997: 37) .

Having been planned and partially built under the auspices of the European colonialists, the city thereby became the Janus-faced child of broke revolutionaries, a fading colonial empire and a network of multinational aluminum companies. MARG built a whole city, with sports infrastructure, such as an Olympic swimming pool (Radio Netherlands Worldwide 2012), football, handball and tennis playgrounds, a judo hall, and athletics facilities (Interview with Activists 15/03/2014); two schools, one for the expats and one that it "borrowed" to the public administration (Pauthier 2002: 74–75); religious buildings, such as mosques and churches (Interview with Former executive Rusal 4/04/2014; Activist 3/04/2014); electricity and water supply that was directly linked with the factory; and many other services, such as waste water treatment, waste disposal, and not to forget a comparably sound health infrastructure with the best hospital in Guinea (Knierzinger 2016: 142). Still in the 1990s, even the president sent his family for treatment to Fria (Interview with Trade unionist 19/01/2014; Former executive Rusal 4/04/2014). Electricity was practically for free for the whole population of Fria. Households had to pay a small allowance to the state-owned EDG, which was responsible for maintenance. However, as only a small fraction of the population actually paid and the little collected money was mostly misapplied, Pechiney de facto took over the management of electricity after a few years (Interview with Journalist 17/01/2014). From the 1950s until 2013, the electricity for the whole city was generated by several thermal power plants installed on the refinery site. Ibrahima Diallo also alluded to the symbolic importance of the factory sirens ordering the everyday life of most of Fria's inhabitants by marking the opening and closing times of the factory as well as signalizing accidents and dangers (Diallo 2014aa). When the sirens were tested 8 months after the halt of the factory, street vendors began to dance on the streets shouting "The whites are back!". However, for good or bad, they had not come back at the time of writing (Knierzinger 2016: 142).

Having constructed neither ward nor school independently, the Guinean state was forced to act as a solicitant and additionally became increasingly indebted to the private company. The finance policy of Pechiney was rather unclear, as it never referred to its public investments as gifts, yet it also never demanded repayment. Such expenditure—called "opérations specials"—comprised buildings, engines and other machines, water, electricity, petrol, maintenance, etc.[23] MARG, being neither private nor officially public after the French retreat, at the same time served as a transmission belt between the ancient colonizers and the newly independent state.[24] Pauthier thus observes "a certain continuity between the colonial and the postcolonial period" (Pauthier 2002: 74–75; translation by JK) in Fria. The break of independence was attenuated by the presence of MARG in Fria and its consequences only came fully into force with the dissolution of MARG in 1960. In 1959, Sékou Touré attached MARG to the *ministère de l'economie générale* and therefore actively claimed control of the organization, which was still financed by the French government. Half a year later, MARG was dissolved (Dulucq et al. 1989: 95–106; cf. Pauthier 2002: 78).[25] The dissolution of MARG made the political consequences of the financial situation in Fria even more difficult, as Pechiney became even more important to the inhabitants. Since the 1970s, Fria invested systematically in public infrastructure that was only of peripheral interest to its economic activities. This concerned among others 16 public schools and several roads (Dubréka-Télimélé and Fria-Katourou; Larrue 1997: 177). Until today, the ownership and the responsibility for public buildings in Fria remain unclear to the population. According to members of the NGO ARSYF, attempts to renovate seemingly public infrastructure, such as the swimming pool and the cinema, had been blocked by the mayor and the prefect by stating that these belonged to the companies (Interview with Activists 15/03/2014; cf. Activist 3/04/2014). Other buildings, such as the *Mosquée centrale,* have been financed by Pechiney and subsequently "donated to the population" (Interview with Activist 3/04/2014). In 2010, Rusal announced the renovation of the same mosque (Deguenon and Traoré 2010: 1). In one of the few Guinean sources on Fria, a government representative already called in 1957 for a state intervention to impede the town from becoming a "concession Pechiney" (Pauthier 2002: 75). According to this source, the company directly distributed the permits for merchants in the shopping arcade and was even directly asked by the churches to construct a "lieu de culte", apartments for priests and other church buildings (Pauthier 2002: 74–88). Based upon similar findings, Soumah concludes that "[i]n fact it is the plant that set up the town as well as the public civil service of Fria" (Soumah 2008: 103).

In the beginning, the state was represented in Fria only by three government officials, a police commissioner and four agents, ergo two superiors, and six employees, against 200 expatriate senior officers on the corporate side (Pauthier 2002: 75). Referring to an audit in the 1980s, Ibrahima Diallo talks about 517 expatriates and 911 "Africans" in 1959. Already in 1963, this relation had changed to 345 expatriates and 978

"Africans" (Diallo 2014f), and in the late 1970s, only a few dozen expatriates were left (Interview with Activists 15/03/2014). My interviewees perceived Fria as "the most cosmopolitan city of the country" (Interview with Journalist 3/04/2014) and as city of "intellectuals" who constituted "somehow a selection of Guinea" (Interview with Former executive Rusal 25/03/2014; Trade unionist 18/01/2014). This is related to the recruitment system of Pechiney in the form of typically French *concours*. The persons who supervised these qualifying examinations were called "psycho-techniciens" (cf. Barry 2008: 22; Journalist 3/04/2014).[26] In 2008, Barry found that only 6% of the core workers were analphabets, 22% had primary education, 58% had secondary education, and 20% were professionals (Barry 2008: 20).

3.1.1.2 Environmental Damage

Pechiney began to act upon ecological concerns much later in Guinea than in metropolitan France. While the ecological consequences of aluminum production (mainly red mud waste from alumina production and fluorine emissions of smelters) were already addressed in France in the 1960s (Boullet 2000), Pechiney only tackled this problem in Edéa and Fria in the 1980s. In the case of Edéa, this did not happen earlier, because the Cameroonian government did not force the company to do so. Since operation costs were estimated to rise 5–8% if more ecological technologies had been applied, the company decided rather to pay "a little compensation" instead of building "costly installations" until it was forced to change policies by the government (Pezet et al. 2009: 33).

In the case of Fria, government officials talk about 30 years of uncontrolled pollution (Interview with Administrator Fria 5/04/2014). A former senior manager recounted that "the water was red down there, but we did not pay anything" (Interview with Former executive Rusal 7/04/2014). The red mud and other wastes of the factory were simply discharged into the Konkouré and other nearby rivers (Interview with Former executive Rusal 25/03/2014) and they continue to pollute these rivers until today (Interview with Administrator Fria 5/04/2014). Fishing in the Konkouré was not possible (Interview with Former executive Rusal 7/04/2014). Fria was not densely populated before the construction of the alumina factory, but all the current villages already existed by that time. Under the Touré government, this local population and the new residents of the city were relocated arbitrarily and hardly received compensations for damage to their property or their health

(Interview with Administrator Fria 5/04/2014; cf. Former executive Rusal 7/04/2014). The first relocation that was combined with documented compensations happened in 1996 in Boudoungoungourou in the sub-prefecture of Baguinet and led to heated debates about corruption. The population did not want to be compensated in kind because they feared that the administration would keep the money for themselves. Pechiney gave in and paid in cash, but as the mine never arrived in this village, the population has not left until today (Interview with Administrator Fria 5/04/2014).

The first environmental code was only voted for in 1986 under the new government of Lansana Conté (ibid). In Fria, the dumping of red mud was restricted for the first time with the creation of Friguia in 1973 (Larrue 1997: 177). However, until today, the ecological standard of the plant remains alarmingly low compared to other alumina factories, such as Nikolaiev (Interview with Former executive Rusal 25/03/2014). In 2006, the caustic soda that had been drained into the Konkouré for 50 years led to a pH of 11 at the point of entrance at a distance of 3–4 km from the factory and only dropped under 8 after 500-m downstream. Across the river at the point of entrance of the red mud lies the village Dotéba. For 300-m downstream, farming on this riverside is not possible, because the soil is too basic (Bah 2006: 30; Engineer 23/01/2014; Journalist 19/01/2014). The same applies for the groundwater around the factory. Even 1 km down from the factory in the *Vallée de Tigué*, the ground water is not consumable and the rivers are polluted. The 2000 inhabitants of this valley were also regularly confronted with accidents in the factory, leading to the drainage of huge amounts of caustic soda. Besides ecologic damages, this has led to numerous cases of alkali burn caused by polluted water. Several other villages face similar problems. The last of these accidents happened 2009 when 19 people had crossed polluted water up to their haunches and saw their skin come off their legs due to that (Bah 2006: 17–28; Engineer 23/01/2014; Diallo 2010: 10; Diallo et al. 2011: 68; Administrator Fria 5/04/2014). Because the dams for red mud are not secured, several children and young men have already lost their lives when trying to walk through or when searching for iron that had been dumped together with the red mud. I have visited the families of two pupils who died in 2006 and 2007 in short tunnels that they had dug in the mud in the search for iron. They had sold the unearthed iron to Chinese and Guinean collectors, according to the family to buy food and clothing (Interview with Street vendor 1/04/2014; Street vendor 5/04/2014).

3.1.1.3 Fria Turns into Friguia

Fria had been called the "Little Paris" of Guinea until the early 1970s, due to its comparatively high salaries (Interview with Journalist 28/03/2014) in combination with extensive fringe benefits and social services (Interview with Trade unionist 19/01/2014), its busy nightlife, and its comparatively broad range of imported consumables (Soumah 2008: 215). However, the radicalization of the Touré regime after an attempted putsch in November 1970 brought this pulsating city life to an abrupt halt. The regime first limited political freedom in town and then willingly brought down the standard of living. Until the end of the 1960s, Touré had already scared away a considerable number of leftist intellectuals, such as Charles Bettelheim or Samir Amin, who had arrived in Guinea full of sympathy and zest for action after independence.[27] Guinea became increasingly isolated, not least due to the fall of the friendly Sierra Leonean, Malian, Ghanaian, and Algerian regimes (Adamolekun 1976: 9; Johnson 1970: 358f). This situation was aggravated by the Portuguese attack on Guinea on 22 November 1970. In the so-called Operation Green Sea, about 400 Portuguese and Portuguese-led Guinean fighters tried to overthrow the Guinean government, capture the resistance leader of Guinea-Bissau, Amílcar Cabral, and free Portuguese prisoners. The Guinean forces offered little resistance, but because neither Cabral nor Touré were to be found the soldiers retreated after having freed their compatriots from the notorious Camp Boiro.[28] Soumah described the following actions by the regime as a "Soviet-like political purge on the entire Guinean territory" (Soumah 2008: 92). The regime murdered dozens of high officials who were called conspirators and more than half of the senior staff of Fria was either imprisoned at Camp Boiro, expulsed or left voluntarily. Larrue reports on the "decentralization" of the imposition of the death penalty to prefectural authorities, which led to the hanging in the city center in 1972 of a "Peul" who had allegedly taken part in the Operation Green Sea. The French expatriates were invited by the regional authorities to watch the execution. Several French families decided to leave after this incident (Larrue 1997: 107, 176). The imprisonment at Camp Boiro concerned both Guineans and expatriates and is remembered as "1 year, 1 month and 1 day" by the workers of Fria (Diallo 2013b).

Another consequence of this radicalization was the foundation of Friguia, a joint venture between the Guinean state and the involved companies, implemented in 1973. According to Soumah, the joint venture was suggested by the private shareholders themselves to avoid being constantly considered as opposed to the regime (Soumah 2008: 92, 110). Given the fact that state involvement was virtually non-existent in global bauxite mining before the 1970s, this remains questionable (cf. Holloway 1988). Sékou Touré also enforced the replacement of the expatriate director of the mine by a Guinean (Diallo 2014g). However, as in the case of CBG, the mixed company still left most important decisions in the hand of the industrials. The CEO of Friguia was proposed by the companies and the intervention of Guinean politicians was limited due to both their access to formal decision-making bodies—such as the board of directors, the *Comité technique,* and the *Comité financier*—and their experience. Most of the public representatives at Friguia had not been trained for their job and did not know which decisions had been taken before their arrival. In addition, the retention of decisions that had been taken started to take place outside of the Guinean territory by 1972 when Frialco—the new holding of the private shareholders of Friguia—shifted their company headquarters to the Caribbean (Larrue 1997: 294–295).

The outcomes of this special marriage of convenience between the—at least rhetorically—revolutionary regime and the global aluminum industry are discussed controversially. While the business and technical elite of the enterprises mostly complained about the loss of effectiveness, the politicization, and the growing nepotism (see Barry 2012), Pauthier also suggests that rather than the state taking over the enterprise, this catalyzed the opposite process of the private enterprise increasingly becoming the public entity itself, mainly due to empty Guinean treasuries (cf. Pauthier 2002: 8–10; Soumah 2008: 105–110). At least some consequences of the increased interest of the socialist regime for Fria since the 1970s were positive: besides a renewed push for the construction of an integrated aluminum chain, the government also enforced the regular employment of subcontractors and the accelerated Africanization of the personnel. Workers also successfully struggled for higher salaries and better working conditions (Larrue 1997: 178, 310).

However, the change of leadership that followed the foundation of Friguia points to a rather destructive influence of the government on the already despotic political reality of the mining town. The ubiquitous brother of Sékou Touré, Ismael Touré, who had already been minister of economic development during the 1960s—where he was also responsible for mining—became president of the board of directors of Friguia (Soumah 2008: 92, 159). Ismael Touré was also the "chief prosecutor" of the Camp Boiro prison, where he was regularly present and actively took part in the torture of prisoners. The German entrepreneur Adolph Marx describes Ismael Touré's atrocious acts of torture in detail (Marx 1977: 9–26, 64, 91–97). During long torture sessions, his torturers—many of them members of the government—accused German firms and "their Nazi networks" of conspiring with the Portuguese. German volunteers were expelled and some of them incarcerated and tortured, including Marx (Marx 1977: 6, 21, 54, 79; see also Barry 2004). By contrast, the US government—which was still about to bargain with Guinea about the reserves in Sangaredi—was left out of the conspiracy theories. By the time of the invasion, the Guinean government was still part of a joint venture with the US company Harvey, although the company struggled to scrape together the necessary money for this huge endeavor. Richard Nixon condemned the invasion and sent a message of "sympathy and support" together with some money for "reconstruction". Touré vehemently criticized NATO, but not its leader, the US. According to a report in the German newspaper, *Der Spiegel*, none of the 159 US expatriates were affected, although some of the US citizens, the employees of Pan American Airways, and the US Peace Corps had already been expelled in 1968 (Der Spiegel 1971).

As a consequence of Operation Green Sea, Touré increased the small military force from 5000 to 13,000 and incorporated it into the party structure. He also began to form a "popular militia" of 300,000 men (Adamolekun 1976: 124, 146). At that time, the rule of the party was already firmly established, with about the 8000 so-called *Pouvoirs révolutionnaires locaux* (PRL) in every village and urban district, which collected taxes, administered "customary law", and maintained "law and order". These units were grouped into 210 sections (with 13 representatives each) and 30 federal bureaus (again with 13 representatives).[29] The paramount unit of the party, the National Political Bureau, started with 17 members after independence. In 1976, after various "conspiracies"[30] against the regime, seven members were left. The others had emigrated or were killed. It remains unclear if these conspiracies have been invented by Touré to get rid of political opponents or not (Schicho 2001: 335; Adamolekun 1976: 16–31).

Workers as Super-Citizens

By 1975 at the latest, the increased presence of the Guinean state in Fria turned against the town's inhabitants. The announcement by the Guinean government to cut fringe benefits (mainly concerning foodstuffs and medical care) in Fria led to a general strike and public upheavals, which were answered by detentions, relocations, and the repression of the trade unions (Barry 2008: 18; Soumah 2008: 106). The socialist government described the strikers of 1975 as "counter-revolutionaries", despite their obvious contribution to class struggle. The reasons for these actions seem to have been mostly populist. Boubacar Bah, a current opposition leader, offers the following account:

"That's socialism: basically, everybody has to be equal. If everybody has the right to 5 kg rice, why has the worker in Fria a right to 15–20 kg? Why, if in Conakry people get sugar and oil, the people in Fria get butter, jam, etc.? Thus, the people of Fria were considered super-citizens. And these super-citizens are dangerous for the revolution" (Interview with Police 28/03/2014; translation by JK).

As the salaries and the fringe benefits were paid by Pechiney and its partners, these workers did not directly harm anyone in Guinea by demanding higher salaries and better working conditions. The taxes paid by the foreign companies were not affected by their expenditure and the Guinean stakes in the mine only paid off in 1989 when the company reported the first "financial surplus from which the government was able to benefit" (cf. Campbell and Clapp 1995: 429; see also République de Guinée 2013c). The workers, nonetheless, seemed to constitute a "discursive" danger as they were fed by "capitalism" and seemed to be quite well off. Apart from that, the government seemed to be afraid of a "contagion" of CBG and used the strike in a similar manner as the various "conspiracies" to get rid of unpleasant trouble makers.[31] In comparison with the public servants, the workers in Fria were not particularly rich. They did not even receive twice the meager salary of their colleagues in the administration. However, remuneration in kind—all imported from abroad—increased their purchasing power considerably.[32] If we consider the hybrid ideological character of the regime, fronts could also have run totally differently: a former senior manager at Fria argued that the government simply sided with Pechiney against the workers (and not with "the Guineans" against Pechiney and its workforce), because it was heavily dependent on taxes from Fria until the start of production of CBG (Interview with Former executive Rusal 3/04/2014).

Sysmin Programs in Fria

In 1980, Fria received a loan from the European Investment Bank and the French *Caisse Centrale de Coopération Economique* within the framework of the first Lomé convention. Despite this investment, the output of Friguia decreased in the following years. Rather than augmenting production, Pechiney used these funds to enhance the quality and reduce production costs, the latter mainly by continuing to exert pressure on the Guinean government (Campbell 1986: 51–53). In 1985/1986, 1 year after the death of Sékou Touré and the putsch of Lansana Conté, the plant received another investment from the European Commission and the EIB to change the type of alumina produced. The current "floury" type, which went along with considerable pollution, was replaced by a "sandy" type of alumina, which further reduced the output. To improve the output again, Fria requested for Sysmin finance, which finally began in 1991. Three-quarters of the connected rehabilitation were budgeted for measures to increase the output from 580,000 to 700,000 tons of alumina per year; another quarter was again destined for reducing costs. The report is quite clear in this respect:

"Friguia, for historical reasons, managed the schools, the city hospital and supplied free water and electricity for the population. This all added to the costs of alumina production and the aim was to remove this responsibility from Friguia. Pechiney, the major foreign shareholder, was responsible for the rehabilitation and the Guinea government was responsible for the peripheral operations" (Ralph Spencer Associates 2000: 49).

Neither the rise in output nor the disengagement from non-industrial activities had been achieved by the Sysmin program (ibid: 81–83). However, Washington Consensus had, nonetheless, arrived in Fria and was welcomed by a general strike of 1 week in 1991 (cf. Barry 2008) and several uprisings in 1992 when workers destroyed the buildings of a newly established company that should collect electricity fees and enforced the evacuation of the prefect of Fria by helicopter. From March to September 1992, the city was without electricity (Larrue 1997: 186). In 1993, Pechiney announced that it lost 15 USD per ton of alumina produced: the sale of one ton brought 165 USD, while the production costs stood at 180 USD. This led to a definitive end of the so-called Economat, the store where workers had been able to buy cheap foreign foodstuffs and other goods. In the 1980s and 1990s, Pechiney profited strongly from French and European development aid for

Guinea. However, in 1997, all investors, nonetheless, sold their stakes to Pechiney. Shortly thereafter, Pechiney handed over the factory to the state for the symbolic sum of one dollar (Diallo 2012g; Sidibé 2012e, g; Storozhenko 2009).

Pechiney Leaves

The interviewees and the literature give various reasons for this degradation of the situation in Fria. Pechiney itself was hit hard by the crisis of the 1970s and was nationalized under the government of Mitterrand in 1982. Outdated equipment of the factory had never been replaced until Pechiney's departure in 1997 (Interview with Journalist 3/04/2014; Larrue 1997: 183, 201). Two years before the unwanted nationalization, Friguia' taxes were again reduced considerably (Dux 2000: 253). According to a trade unionist, the departure of Pechiney was finally triggered by a change in the administration. A group of young technocrats around the Prime Minister Sidya Touré (1996–1999) resolutely turned down Pechiney's continuous demands for cuts in social spending in Fria and thereby prompted it to finally carry out its threats (Interview with Trade unionist 19/01/2014). Others underline the role of the trade unions. Ibrahima Diallo, a former stationmaster and influential intellectual in Fria who normally sides with the workers against capital, talks about "disproportional ambitions" of the former (Brygo 2009; Diallo 2013k). Several newspaper articles also blame the mismanagement by the state since the joint venture of 1973 (e.g., Guinéenews 2015; see also Dux 2000: 252). While all these things contributed to the decline of Fria, the most important reason was without doubt the fall of aluminum prices during the 1990s, which led to the downfall of most of the Six Sisters. Between 2000 and 2003, Alcoa bought Reynolds, Rio Tinto acquired Comalco, Norsk Hydro bought VAW, and Alcan bought Alusuisse and Pechiney. In addition, three new global aluminum players emerged: Rusal, Chinalco, and BHP Billiton (see Fig. 1.5). Friguia was far from being the only plant that operated at a loss at that time.

When the state took over the 51% left from Pechiney, it also assumed a debt of 130 million USD, much of which had been accumulated during the end of the 1990s and was now even more difficult to handle as several Western banks retreated with the Western operators. Fria was thereby forced to accept insolvency proceedings and started a process of reprivatization under the auspices of the World Bank. Different evaluations estimated the value of the company between one USD and

20 million USD. A first bidding did not lead to any results, mainly due to the world economic situation (cf. Dux 2000: 254; Storozhenko 2009; Sidibé 2012e). From 1997 until 2000, Fria was thus fully state-owned. With the exception of two expatriates, it was exclusively run by Guineans (Diallo 2012g, 2014f). Ibrahima Diallo describes this time as a success story in technical terms, only troubled by the uncooperativeness of certain furnishers and by corruption of certain representatives of the government. The handover of the company town in 2000 to Reynolds accordingly amounted to him to the self-abandonment of the Guinean government. Reynolds was also in the red and was bought by Alcoa in the same year when it signed its management lease contract with the Guinean state.[33] The rent for leasing the facilities was directly transferred to the former loaners.[34] As Pechiney before, Reynolds tried to hand over the management of water and electricity to public companies, but it also failed. Moreover, as Pechiney before, the Americans had plans to erect an integrated aluminum industry in Guinea. Ibrahima Diallo possesses the picture of a signboard that was erected in the vicinity of the factory, stating: "Télémélé, Tabossy, usine d'aluminium, superficie 60 ha" (Diallo 2014f). As Reynolds only employed 22 expatriates, the dominance of Guineans in the company was retained. When it handed the firm over to Rusal in 2003, the number of expatriates rose dramatically (Barry 2008; Sidibé 2012e).

Many informants saw in the 1990s the definitive end of a modernist dream, which admittedly had already shown some cracks in the mid-1970s. With the end of the *Economat* in 1996, purchasing power slumped considerably, while "public" infrastructure was no longer maintained and social services seriously degraded (cf. Trade unionist 19/01/2014; Engineer Rusal 2/04/2014). A trade unionist summed up that the worker in Fria had been "really a fellow who was at ease. [...] Thus, you have seen where we are coming from. We come from far away, to fall so deep" (Interview with Trade unionist 19/01/2014; translation by JK). A sociologist at GAC described this economic downfall as a mass psychological crisis that tore apart families and even became part of the Guinean mentality. She grew up as the daughter of the superintendent of the municipal swimming pool in Fria and described her childhood as a paradise lost. When the American junior company GAC (see below) began to draw a similarly colorful picture of social progress in the prefecture of Boké, many *Friakas* adopted this project as their own. However, the sub-prime crisis also brought those dreams to a sudden end (Interview with Sociologist GAC 1/02/2014; Trade unionist 18/01/2014).

During my stay in Fria, I encountered the family of a welder, whose fate was exemplary of many others. His wife grew up in Fria with both parents working at the factory and spoke exceptionally good French. In 1998, during the crisis, her husband started to work for the Guinean subcontractor ACM. By then, he received 150,000 GNF (about 140 USD)[35] per month without primes, without medical care or other fringe benefits and without paid overtime. Only once or twice per week, he worked 8 h per day as provided in the contract, whereas on the other days, he worked up to 16 h. Only through the pressure applied by senior employees of Friguia on the owner of the sub-contracting company, his wage was raised to 700,000 GNF (about 420 USD) in 2000 and 1,600,000 GNF (about 360 USD) in 2006. With the takeover of Rusal, his salary slumped again to 980,000 GNF (220 USD). In 2008, shortly after their marriage, rather than enjoying a honeymoon, the welder was forced to work for 3 weeks without break from 7 am to 11 pm, with the promise that he would be employed permanently shortly after. Until the lockout in 2012, this had not happened. By the time of the interview, he was jobless and received assistance from friends and relatives to ensure the schooling of his five children in private schools. The school fees alone amounted to 2.5 million GNF (about 360 USD) per year (Interview with Worker Rusal 3/04/2014).

3.1.1.4 Fria Under Rusal (2003–2014)

In March 2003, Rusal became manager and tenant of 85% of Friguia, which had been held by Reynolds before, while the rest remained state owned (Diallo 2013e). According to Rusal, the alumina from Fria was already exclusively shipped to Russia before the company acquired the management lease from Reynolds in March 2003 (Storozhenko 2009). To the population and apparently also to parts of the government, the takeover of the management lease came as a surprise (Diallo 2010: 7, 2013e; Guinéenews 2015). In September 2003, Rusal and the Guinean government started negotiations on the privatization of the factory. An audit effectuated by Ernst and Young estimated the value of the facilities in a range between 22 and 210 million USD. A following call for bids—which was only partly made public—was won by Rusal. In March 2006, the company became the sole owner of the plant by (1) paying 22 million USD directly in the Guinean treasury, (2) taking over 105 million

USD in debts, (3) nullifying 5.5 million USD of debts of the Guinean government vis-à-vis Rusal that had been contracted by Lansana Conté to buy 20,000 tons of rice to calm the population of Conakry, (4) investing 5 million USD for the asphalting of the streets of Fria and 15 million USD for a fifth power generator, and (5) promising to invest 650 million USD in the near future to more than double the alumina production (Guinéenews 2015). It also guaranteed the supply of water and energy for the population of Fria and the maintenance of the infrastructure. The conditions of this privatization are highly contested until today.[36] After the takeover, the Guinean newspaper Horoya reported:

"The extension and the modernization of the factory ACG/Friguia proposed by Rusal will enable the installation of modern and environmentally safe production, new jobs for the population and more value added for both parties (Guinean state and Rusal)" (Camara 2006; quoted in Diallo 2012c; translation by JK).

In April 2006, a delegation of the trade unions went to the mining ministry to obtain more information about the privatization, but "they came back in vain" (Diallo 2013e; translation by JK). One of the competing projects in the bid was a joint venture between the Greek company Third Party Logistics, the China Aeronautic Technology and Industry Corporation (CATIC), and Chinalco, which proposed to buy the factory, build a hydro-electric dam on the Konkouré River, and construct an aluminum smelter and a port in the Boffa prefecture. However, the investors did not want to buy (and subsequently maintain) the *cité* and the railway.[37]

Like almost all of its predecessors in Guinea, Rusal won the bid with the promise of an extension project: the production of 600,000 tons of alumina should have been augmented to 1.4 million tons and the company intended to erect a so-called "autoclave", a pressure chamber that would have limited the loss of alumina during production (Interview with Former executive Rusal 25/03/2014). Claiming that the Guinean government was reluctant to approve these plans as provided by law, Rusal has not implemented any of these measures until the lockout in 2012 (cf. Storozhenko 2009). Interviewed workers in Fria did not agree with this interpretation of Rusal: due to the poor condition of the factory and owing to the fact that the company preferred to process its bauxite in the Ukraine, Rusal arrived in Fria with the aim of reducing it to a bauxite mine. According to Ibrahima Diallo, this was expressed openly by the company in 2003 when it took over the management from Reynolds. The Guinean government, nonetheless, forced the company to maintain the factory, which it did quite halfheartedly (Interview with Trade unionist 19/01/2014; Diallo 2014f; Mayor of Fria 18/01/2014). Despite its announcement to more than double production when Rusal bought Fria, the company only maintained power generation for the city and bought six new locomotives (Knierzinger 2016: 143f; Deguenon and Traoré 2010: 1; cf. Journalist 19/01/2014; Journalist 15/01/2014).

The New Masters

Despite the rather humble overall performance of Pechiney in Guinea and the continuous disputes and conflicts with government and workers, the French company is highly esteemed among the *Friakas* today and seems to be an important element of a profound local patriotism (Brygo 2009; Trade unionist 25/03/2014; NGO Reunion 29/03/2014). The respective statements in the interviews were often part of comparisons with the current proceedings of Rusal and contained the term "culture" in various cases:

"Corporate life where they come from is totally different to here. Here syndicalism has enabled the workers to make demands. When there is a problem, syndicalism enables to address the problem. But my acquaintances have made me believe that this is a new element where they come from. The worker does not have the right to make demands. This friction, I think that this was very difficult for them [...] Thus it's true that there is a difference in the working life between what they are used to and what they have found here" (Interview with Mayor of Fria 18/01/2014; translation by JK).

Even the Guinean workers who were on formation trips in Russia seemed to have caused problems, because "this contact widely opened certain eyes over there in terms of work relations" (Interview with Mayor of Fria 18/01/2014; translation by JK; cf.Soloviev 2011). Moustafa N'Diaye, *secrétaire administratif* of the CNTG, remarked that Rusal had no experience with capitalist production when it arrived in Guinea:

"That is why sometimes they use a coercive approach, sometimes they are a bit social and sometimes they act like capitalists, but with a rigor we do not understand. Because these people, they do not have this culture" (Interview with Trade unionist 28/03/2014; translation by JK).

A journalist and former employee of Friguia equally described the new leaders as "traditional chiefs" who replaced a good working atmosphere by coercive power. Moreover, Ibrahima Diallo underlines the openness of French and American employees versus the Guinean culture in comparison with the attitude of Rusal, which erected a "Berlin Wall" (Diallo 2014g) in the city after its arrival:

"They arrived and first of all they preferred to reduce contact with the population. They had heard that the French whites lived in the same habitat with the Guineans and shared their meals and that the Americans did almost the same. But they did not want to do that. Thus, they erected a *cité* right in city center that had privileged access to all the existing infrastructure and an extension of the initial *cité*. The principal characteristics of these two entities were the fact that they were closed (3-m walls with barbwire) and that they could only be entered by Russian personnel" (Diallo 2014f; translation by JK).[38]

The so-called "Small Moscow" was created in 2006 by merging several existing villas and endowing them with two diesel engines and a well drilling (Diallo 2013e). Most interviewees associate the takeover of Rusal with a further degradation of living conditions in Fria, a profound management crisis (Interview with Trade unionist 18/01/2014; Journalist 15/01/2014) and the rise of favoritism, notably concerning the recruitment of personnel (Barry 2008: 22; Journalist 3/04/2014), the award of contracts (Diallo 2014g), the granting of loans by the company for construction, and other investments of the workers through the so-called "solidarity fund" (Barry 2008: 35–36). One interview partner recounted a case, where a candidate paid 8 million GNF (about 1000 USD) to get a job in Fria (Interview with Administrator Fria 5/04/2014). The management of Rusal was perceived as less intelligible, less trustful, and less elaborated than one of its predecessors:

"When it arrived, Rusal disapproved what it saw in Fria: Russian cadres who were used to live together in settlements and to share transport, found themselves living next to blacks whose opulence was superior to theirs. And the greatest provocation was that the agents in the factory had the necessary competences to run the installations, despite the skills of the competent cadres and the great experiences of the Russians who were periodically replaced and who could not digest to be ridiculed by the foremen. We think that it was in Fria that the personnel of Rusal became aware of the gap that separated them from the so-called capitalist development. They did not like that" (Diallo 2012k; translation by JK).

By gradually introducing a parallel management structure controlled by Russian nationals, Rusal started de facto a process of deafricanization and thereby also lost the loyalty of the Guinean seniors who were left (cf. Diallo 2014g, 2012c; Journalist 15/01/2014). Having been 100% Guinean after the retreat of Pechiney, the factory received more than 60 expats—mainly from Russia—with the takeover of Rusal. Unequal payment on racist terms that had already preoccupied trade unions throughout the history of Fria[39] now seemed to retransform into patterns of colonial apartheid with the important difference that many Guinean workers now described the new European employees not only as disrespectful but also as incompetent (cf. Journalist 15/01/2014). Most of the Russians spoke neither French nor English (cf. Diallo 2013e).[40]

Sub-Contracting

The introduction of new sub-contracting firms was probably the measure with the most negative implications for the *Friakas*. Rusal started this process after its takeover in 2006, but faced considerable resistance by the Guinean workers (Diallo 2013e, 14/03/2006).[41] In 2007, the company, nonetheless, succeeded in pushing back Guinean providers—which were often directed by former workers (Interview with Engineer Rusal 4/04/2014)—and replaced them with a complicated interwoven network of Russian-led sub-contracting firms, led by EMG (*Entretien et maintenance en Guinée*), EMF (*Entretien et Maintenance Friguia*), and CGG (*Compagnie Guinéenne de Génie*). These companies soon found themselves on the apex of a wide network of subcontractors. According to Rusal, this network comprised 199 local enterprises in 2008 (Storozhenko 2009; cf. Former executive Rusal 4/04/2014). CGG apparently already existed before Rusal, but was renamed EMG in September 2007. It employed 400–700 workers,[42] was responsible for the administrative management of the other companies, and acted as a service provider in the fields of human resources, the management of housing, relations with public authorities, and certain construction works (Diallo 2010: 29; cf. Storozhenko 2009). It thereby commanded over a wide range of further subcontractors (Interview with Mechanic 4/04/2014) and constituted a parallel hierarchy within the factory that increasingly took over decisions of the Guinean core personnel (Diallo 2012c, 2013e). The old CGG had been owned by a Guinean who has apparently been forced out of business. According to the Guinean owner of another sub-contracting company, Rusal had enforced higher wages while paying lower commission. When the owners gave up, EMF took over the workers (Interview with Entrepreneur 7/04/2014; cf.

Mechanic 4/04/2014). Ibrahima Diallo also accused Rusal of not paying numerous bills to Guinean subcontractors or for deliberately paying them up to 3 months too late (Diallo 2012k; Entrepreneur 7/04/2014). The EMF had various tasks, including human resources and technical maintenance. The CGG was mainly providing construction engineering for the two others and hired local contractors for engineering. Apart from EMF, all the other subcontractors additionally shared one common directorate that also extended to the subcontractor EMC in Débélé. Their services are based upon agency agreements or service provision arrangements coordinated in Moscow (Diallo 2010: 29). This complicated system had apparently been reshuffled several times (Interview with Former executive Rusal 7/04/2014; cf. Diallo 2010: 29). A subcontractor divided these firms into three categories. The EMF had a privileged status between contractant and subsidiary, whereby workers received better wages and had certain fringe benefits. The other mentioned companies received renewable contracts, with the arrival of Rusal every year, since 2010 every 6 months and later even every 4 months. This concerned about 1500–2000 workers. A third category of about 2000 workers had no contracts at all and made bids on their own. The interviewee had a network of about 50 masons, electricians, painters, etc. that he hired on this basis. This adds up to a total workforce of 5000–6000, with one-third of these workers only being hired occasionally.[43] Despite becoming increasingly important for the functioning of the factory, categories 2 and 3 received salaries below living costs, starting at less than 3 USD per day for day laborers and about 40 USD per month for contract workers (Interview with Journalist 3/04/2014; Activist 3/04/2014; Entrepreneur 7/04/2014). Fringe benefits varied considerably from one firm to the next and also seemed to be highly dependent on personal ties with decision makers. Aside from the employees of EMF, most workers did not receive medical care and were not accommodated (Interview with Former executive Rusal 4/04/2014; Mechanic 4/04/2014; Journalist 25/03/2014). Subcontracting already existed under Pechiney (Interview with Journalist 25/03/2014; Former executive Rusal 7/04/2014), but only seemed to have become important during the short phase of nationalized production from 1997 to 2000. In 1971, Pechiney had in fact been forced by the government to regularize many contract workers (Larrue 1997: 178). In many cases, old workers who were threatened with retirement founded small sub-contracting companies. Rusal then used this structure to establish its own hierarchy within the factory and further reduce costs (Diallo 2012c, k; Engineer Rusal 4/04/2014).

More Conflicts

Based upon these developments, conflicts in Fria increased. Already in December 2004, the population protested, because Rusal did not respect the contract conditions. Clashes with the police resulted in one death, 18 serious injuries, and more than 100 detentions (Diallo 2010). In 2005, the lights went out in Fria, because a generator stopped working. This necessitated the total halt of the factory and thereby stopped the provision of electricity and water for the whole city (Diallo 2005d). The inhabitants had to walk several kilometers to get water. In the following years, power cuts became frequent (Diallo 2013e; Kadyrov 2010: 3; Journalist 19/01/2014). In March 2006, the personnel of the hospital protested, because there were not enough medicaments (Diallo 2013e). This shortage was probably related to struggles between government officials and Rusal over the control of the import of medicaments. In November 2011, when the personnel again decried a shortage of medicaments and demanded the demission of the hospital's director, Rusal stated that costs could be reduced by 25–30% if the government would not force Rusal to buy the medicaments locally. It also announced to stronger control the use of the medicaments by its workers.[44] According to a trade unionist, the company had reduced the budget of the hospital in the 2000s by two-thirds, from 1 million to 350,000 USD (Interview with Trade unionist 19/01/2014).

The situation in Fria remained tense throughout 2006. In June of that year, the workers took part in a nationwide general strike (Diallo 2013e). In December 2006, new riots resulted in six injuries. On this occasion, the public administration of Fria was accused of having misappropriated 5 million USD that Rusal had paid for roadwork, in accordance with the contract of 2006 (Diallo 2010; Journalist 15/01/2014). According to figures provided by the journalist Akoumba Diallo, the taxes paid by Friguia more than halved between 2003 and 2006, while sales increased by about 70% (cf. Diallo 2010: 7). During the upheavals, governmental structures (local and national) and corporate governance of Rusal were often perceived as one entity by the population. Protests against Rusal always went along with the destruction of public good and accusations against government officials. In many cases, these ties between local government and the corporations are quite obvious. In January 2014, a journalist described these entanglements as follows:

"Sometimes the people are embarrassed because they think that the prefect is a beggar. [...] Every time when there is a small problem at the factory, the prefect will come and side with the director of the factory. People don't like that. They say that this prefect and all the others are *sold*. [...] Sometimes even the fuel that the prefect uses for his car: He drives to the factory, the people there give him what he wants. That's not normal. He is stealing. He is about to impose himself. [W]hen there are strikes, for instance, the workers listen rarely to the prefects. They have to send soldiers and policemen. They don't listen to him, because they say: no, you are a bandit, you live *in the pocket* of the managers of the factory. That's it. You live in their pocket and that's why you can't say the truth. You are there to defend the Guinean state and all that, but that's not what you do!" (Interview with Journalist 17/01/2014; translation by JK).

The most important reason for these practices is the strong difference in income between workers and administrators. A former trade union representative even stated that the average salary in the factory was higher than that of the prefect (Interview with Trade unionist 18/01/2014). The mayor of Fria had been employed by the factory before and stated that his current allowance as mayor was about one-tenth of his former salary (Interview with Mayor of Fria 18/01/2014). The salaries of several Guinean senior officials reached about 12 million GNF (about 1750 USD; Mayor of Fria 18/01/2014), while the Guinean mining minister stated publicly in March 2014 that he earned only 10 million GNF (about 1500 USD) per month (Interview with Consultant 17/03/2014). Expatriate salaries are at least four times higher than these Guinean top wages, depending on their country of origin (Interview with Police 28/03/2014).[45] A journalist illustrated this underlying power differential by recounting several conflictual situations in meetings between Russian expatriates and governmental representatives, where the lack of respect of the Russian workers, engineers, and managers for their begging governmental superiors became manifest (Interview with Journalist 17/01/2014).

Protests on the Countryside: The Example of Kondekhore

The conflicts in the city were accompanied by protests and roadblocks by the peasants around the mine, due to insufficient compensation for relocations and damages or disagreements within the communities (Interview with Elders Kondekhore 4/04/2014; Administrator Fria 5/04/2014). In many cases, the money did not find its way to the victims (Interview with Journalist 3/04/2014; Former executive Rusal 4/04/2014), villagers were unwilling to leave their homes after having been paid for their houses and fields, and construction works were not finished or not accepted by the population (Interview with Journalist 17/01/2014; Administrator Fria 5/04/2014). The *directeur préfectoral de l'environnement des eaux et forêts* listed an array of unresolved cases, some of which have already been pending since 2006. This concerns air pollution by alumina dust (Interview with Engineer Rusal 4/04/2014; Former executive Rusal 25/03/2014), pollution of the river Konkouré and other rivers, destruction of fields and plantations, numerous injuries, and several deaths (Interview with Administrator Fria 5/04/2014; cf. Former executive Rusal 4/04/2014). In the case of a fatal truck accident, the prefect apparently insisted upon finding an amicable solution with Friguia at the local level rather than remitting it to the courts. An administrator recounted several meetings, where the compensation for the dead man was discussed like the "price of a sheep" (Interview with Administrator Fria 5/04/2014). In many other cases, the local administration simply lacked the necessary funding and know-how to give a sufficient proof to the attorneys working for Rusal. This was often the case with the air and water pollution. In these and other cases, the officers additionally have problems in reaching the areas of complaint. While a great number of the heavily criticized security forces (e.g., Journalist 17/01/2014) are provided with cars, fuel, and accommodation, civil servants often do not even dispose of a motorbike to conduct on-site inspections (Interview with Administrator Fria B 30/03/2014; Administrator Fria 5/04/2014).

One of the most implicated communities around the mines of Fria is the village of Kondekhore, with more than 600 inhabitants (Interview with Elders Kondekhore 4/04/2014). Since the 1950s, the groundwater has been polluted through nearby waste heaps made up of red mud and other by-products. This practice changed in the 1980s, although the caustic soda is washed out of these heaps in the rainy season and thereby kills the fish in the creeks, obstructs tillage, and renders the groundwater undrinkable. The nearby blasting chased away the game, destroyed houses, and redirected the ground water, which resulted in drying up all the three of the nearby creeks (see Fig. 3.2).[46] Moreover, the bauxite

Fig. 3.2 Inhabitants of Kondekhore in a dry riverbed, April 2014 (JK)

trucks continually dispersed dust, which additionally reduced the agricultural productivity (Interview with Elders Kondekhore 4/04/2014; Administrator Fria 5/04/2014). When I arrived in Kondekhore, the inhabitants were initially reluctant to talk about their situation, because the village had already been visited by several groups of experts, first sent by Pechiney, then by Reynolds, and subsequently by Rusal. On each occasion, the inhabitants had to explain their situation and the researchers counted their trees, houses, etc. Every time, these experts concluded that the village had to be relocated, although none of the companies has done so to date. The last study in 2011 took 18 days and did not lead to anything because Rusal left Fria shortly thereafter (Interview with Elders Kondekhore 4/04/2014).

According to the elders of Kondekhore, all the villagers have received to date was a water drilling and 7 million GNF per year from 2000 to 2004 (about 3500 USD in 2004) as compensation for destroyed market gardens after the women of the village had stopped the bauxite trucks (Interview with Elders Kondekhore 4/04/2014). In contrast to these statements, I was then told by the prefecture that the village had been compensated numerous times and also profited from a "Protocole d'accord" between the leaders of Kondekhore and Rusal that stipulated a lump sum of 15 million GNF (about 3000 USD in 2009) per year for all the infractions listed above (Préfecture de Fria 23/10/2009; Administrator Fria 5/04/2014) and obliged

"all neighboring communities that signed the protocol to do everything they can to preserve and guarantee peace, calm and a climate of perfect understanding. They also campaign against all sorts of violence and every act that could affect mining or the functioning of the factory" (Préfecture de Fria 23/10/2009; translation by JK).

I decided to make a case study out of this disaccord and tried to track down these payments of 3000 USD per year in exchange for social peace in Kondekhore. After 1 day of alternating accusations by the involved parties, I had to stop this inquiry without any proof that the money had been paid. Rusal had left its books in the locked factory, the prefectural persons in charge stated that they have forwarded the money, but had no receipts, and the man who apparently took it on behalf of the community was not in town.[47]

Contrary to all other interviewees, the respondents in Kondekhore were rather pleased about the stop of the factory. No inhabitant of Kondekhore had been employed by Rusal and only few agricultural products, such as oranges, grapefruits, cola nuts, and bananas, had been sold in town. Other products, such as mangos and palm oil, had already been consumed by the villagers themselves. With the end of nearby mining, the quality of the water improved, the fruit harvest increased again due to less dust on the leaves, the blastings stopped, and the population was able to use the old and much more direct routes again to reach Fria and the nearby villages (Interview with Elders Kondekhore 4/04/2014).

Dadis and the World Economic Crisis

The situation of conflict in the early 2000s was deepened by the democratic movement against Lansana Conté that amplified during 2007 and the financial crisis that cut the aluminum price by half in July 2008. The slump in commodity prices was felt in almost all metal markets, although the aluminum industry was hit particularly hard, because its production is highly energy intensive and energy prices recovered faster than aluminum prices.[48] The overall dependence of Fria on fuel reinforced this tendency. According to Rusal, fuel made up of 61% of the production costs in Fria shortly before the halt of the factory (Vassiliev 2012: 6). In an interview, a former accountant of Fria put this figure again 10% higher (Interview with Former executive Rusal 4/04/2014). In addition, Rusal had expanded heavily in the years before the sub-prime crisis. After a short window of ameliorations in the global aluminum market from September 2003 until July 2008 (cf. London Metal Exchange Limited 2014), Rusal thus had a whole world economy of reasons to further reduce costs in Fria rather than augmenting production as provided by the contract. The company identified Fria as the least profitable of Rusal's eight alumina refineries at that time and announced vast cost-cutting measures, but, nonetheless, promised that neither water supply nor electricity would be cut, "because the survival of the population" (Vassiliev 2012: 6) depended on the functioning of the factory (cf. Brygo 2009; Diallo 2014f). When asked whether the Guinean government was in support of the actions of Rusal, its representative complained that the economic crisis was not yet seen as an immediate threat by Guineans:

"The crisis did neither come upon the global markets, nor upon the London Stock Exchange, and even less upon the other European trade centers. The crisis came upon the production facilities. The crisis is nowhere else, it is in Fria. [...] It's like in a family: there is only little money left and other funds are not expected to come soon; economizing therefore seems to be a level-headed policy. [...] We guarantee that the enterprise will not be shut down and that the town will be given water and electricity. The workers won't lose their jobs. In the current context, such guarantees are precious. In view of the above, we think that our generous promises and our good intentions concerning Fria will be appreciated by the Guinean authorities" (Vassiliev 2012: 6; translation by JK).

However, the crisis in Fria had already started before the world economic crisis found its way into Guinea. On 26 February 2008, 1 year after the nomination of Lansana Kouyaté as prime minister, the *Friakas* organized a demonstration in support of Kouyaté that transcended into an "excess night" (Diallo 2013e). While the Russian personnel was hiding in the protected compound, the population on the streets shouted slogans like "Down with Rusal!" or "Rusal go home!" Many of the demonstrating women were dressed in red (Diallo 2013e). These protests recurred in March, were also televised and then repeatedly discussed by the government (Brygo 2009). On 2 June, the whole factory was again on strike for a better payment and working conditions. This heated atmosphere was again fueled by the coming into the power of Moussa Dadis Camara in December 2008 (Diallo 2013e, 2014d, f).

In a letter to the mining minister in March 2009, Rusal nevertheless announced further reductions of production and fringe costs and the discharge or retirement of one-fifth of the core personnel (Diallo 2013e, 2014d, f). This provoked a general strike in Fria from 1 to 8 April, demanding better working conditions, the dismissal of the director of the factory Vladimir Kryuchkov and the recuperation (*récupération*) of the factory (Diallo 2013e). Kryuchkov and his Russian colleagues had to flee from Fria (Guinéenews 2015). At the same time, on 31 March 2009, the government banned all subcontractors (EMF, CGG, EMG, BIEX, and RIC) in Fria. This ban lasted until 4 May 2009. On 8 April, Capitaine Camara harassed the director of Rusal Guinea, Anatoly Patchenko, in a broadcast called "Dadis Show", calling him an international crook (*escroc international*) and screaming at him in an embarrassingly meaningless and violent manner. He thereby provoked a diplomatic crisis with Russia (Diallo 2013e, 2014f; Jeune Afrique 2009). This went along with a trial against Rusal for having paid too little for the facilities in Fria and for not meeting the obligations stipulated in the contract. Rusal reacted on 30 April by publishing an "Aide-mémoire sur la privatisation de Friguia", where it tried to rectify information on the privatization and its course of action, arguing (as Pechiney before) that the production price in Fria was far above the price of sale. In addition, the company listed all "social projects" that had been realized until 2009. This comprised the maintenance of water and electricity infrastructure (while the company continued to complain that the population did not pay the allowances; cf. Kadyrov 2010), the renovation of two stadiums, several religious places and the *Monument aux martyrs de la République*, the construction of a center for industrial training, the installation of several well drillings, and the granting of 100 stipends for schooling in Russia (Storozhenko 2009). Several interviewees criticized this list by pointing out that many of these "social projects" were part of Rusal's obligations, whereas several others did not work out, have never been finished, or were part of corrupt deals. This concerned the stadiums, the well drillings (due to a lack of spare parts, only the one of Little Moscow seems to be operative until today), the industrial training center, a mosque, and a church (Interview with Former executive Rusal

4/04/2014; Diallo 2013f; Administrator Fria 5/04/2014). Until its departure in 2012, Rusal also asphalted 15 km of roads and renovated parts of the prefectural service habitat (being responsible for land issues), a youth center, and the police commissioner's office (Interview with Former executive Rusal 4/04/2014; Deguenon and Traoré 2010: 1).

The relations with the security forces were further strengthened before and during the crisis. In addition to proper security forces and the so-called *conseiller militaire*—a military officer who received one wage from the state and another one from the company—Rusal requested 29 gendarmes who received *primes* (boni), meals, and accommodation from the company and were provided with cars. The position of the *conseiller militaire* had been established in all mines of Guinea under General Lansana Conté. Beside their doubled remuneration, they receive accommodation and a car from the companies (Interview with Journalist 17/01/2014). According to a former public relations manager of Friguia, these paramilitary forces often serve to break strikes: "during the strikes these are the guys who come to crack down on the workers and [...] nick them. That's the reality! And they themselves are stealing" (Interview with Journalist 17/01/2014; translation by JK; for general information on the militarization of Guinean mines see Diallo et al. 2011: 41). The allegation that gendarmes and other security forces were implicated in thefts in the factory has been brought forth by several interviewees (see below). According to the same informant, the police commissioner was also close to the factory management (Interview with Journalist 17/01/2014). In June 2013, a radio announced that Rusal had bought two tanker trucks for the police without telling the authorities (Diallo 2014r; cf. Mayor of Fria 18/01/2014).

On 13 May 2009, the director of the refinery, Vladimir Kryuchkov, was reinstalled in Fria by the Guinean military. Two days later, Anatoly Patchenko was allowed to leave Conakry (Diallo 2010: 30). On this day, the mining minister, Mamadou Thiam—an outspoken opponent of Rusal—was replaced by Capitaine Moussa Tiegboro (Diallo 2013e). Since then, the actions of the government became contradictory. On 5 August, Camara apparently met Deripaska in Conakry (Diallo 2010: 31) and is reported to have stated afterwards that the government should "abandon the trial against Rusal" (Diallo 2010: 33). However, shortly thereafter, the president is reported to have been informed that he had not met Deripaska himself but rather a double. Dadis thus changed his opinion concerning the trial again (ibid). This farce probably originated out of disagreements within the government itself, in particular, between the president and his vice Sékouba Konaté, who became president himself after the assassination attempt on Camara in December 2009. Two successive court rulings on 10 August and 10 September finally nullified the privatization of 2006 and ordered an audit to estimate the losses of the Guinean state since 2006. Rusal did not accept the ruling by pointing to provisions in the contract that indicated the *Chambre de commerce international* in Paris as the valid arbitrator in case of disagreements (cf. ibid: 36). In addition, it stopped tax payments and stated that it could replace any loss of Guinean alumina production "with its own production from other facilities or through market purchases" (Burgis et al. 2010; Deutsche Welle 5/2013).

At this point in time, the workers' struggle had already profoundly marked the city and the factory. Having been unsuccessful in reducing direct charges in Fria, Rusal had cut investment in maintenance to a minimum. According to a former quality assurance manager at that time, Fria was already running on outdated technology when Rusal took over (Interview with Journalist 15/01/2014). The permanent situation of conflict since then enforced this course of action. A report of the trade union, entitled "La politique de 'après moi le déluge' pratiquée par Rusal à Friguia" (The politics of "after me, the deluge" practiced by Rusal in Friguia), decried this method in August 2009. The workers claimed that—already since 2006—damaged vehicles had not been replaced and replacement parts for machines had to be obtained by dismantling other machines or were replaced by used material bought in Eastern Europe.[49] This also led to a degradation of security standards, the frequent shutdown of parts of the factory, and environmental damage, such as the leakage of large quantities of sodium hydroxide into the nearby rivers in July 2009. To hold back this leaked sodium hydroxide, Rusal erected a dam that broke again in August 2009, leading to even more environmental damage (Diallo 2012c; cf. Knierzinger 2016: 148). Already in May 2009, Rusal admitted a shortfall of replacement parts itself and explained this with the allegedly empty pockets of Rusal since the financial crisis:

"The delay of delivery of the replacements parts is mainly due to the financial crisis 2009. The consequences of this crisis are still weighting down on us. At the moment, we are using our reserves. In 2009, the difficulties connected to order fulfillment and payment have resulted in delays of delivery" (Kadyrov 2010: 3; translation by JK).

In November 2009, Genady Ulyanich—a Ukrainian and the company's head of communications responsible for the company newspaper Voice of Fria—told Le Monde diplomatique that "[i]n Moscow they don't realize that most workers here are fathers with dozens of mouths to feed... The locals have told me that when they drive us out of here, I'll be the only one they take pity on" (Brygo 2009). He believed that Chinese companies would take over after the retreat of Rusal: "They'll tear everything down and build a new, more modern factory. It would cost too much to renovate this factory" (ibid). The rapid degradation of both the facilities and the city in combination with the low price that Rusal had paid for obtaining the factory led to the general opinion that the company profited heavily from the misery in Fria. The mayor of Fria believes that "in less than a year you can gain back everything you invested. [...] This is why I would say that this [the factory; note by JK] was a present" (Interview with Mayor of Fria 18/01/2014; translation by JK).

Pensions

In December 2009, Rusal unilaterally retired 132 workers. After 3 years of continuous struggle, they finally received 19 months of salaries (Diallo 2012d), but did not have access to the public pension system (Diallo 2014f). After Rusal had announced their retirements, they had to be prevented from entering the factory by military forces (Commission des 132 agents de Rusal Friguia 8/12/2012a; cf. Diallo 2014ab). These potential retirees had been "accumulated" since 2006 when Rusal had suspended retirements, because the public pension funds were empty and the government had been forced by the trade union movement to find a solution (Diallo 2014y; Storozhenko 2009). Already in April 2009, Rusal had announced that it would have to retire these employees in the near future to employ young workers (Storozhenko 2009) and raise the wages of the rest of the personnel (Diallo 2012l, 2014g). According to Ibrahima Diallo, who was one of these 132, Rusal did not replace these workers as announced (Diallo 2014ab).[50] The 132—as they were called from then onwards—thereby fell in between the systems: they lost all their rights in Fria (particularly housing and medical care), but were also not taken over by the public pension system, the CNSS. Both the CNSS and the 132 based their claims upon a decision of the national assembly in 2006 (L/2008/003/AN) to raise the age of retirement (Commission des 132 agents de Rusal Friguia 8/12/2012a; Diallo 2014ab) and the

CNSS additionally complained that Rusal had not paid its contribution to the fund (Diallo 2014f). Rusal did not accept this law, stating that it only concerned civil servants (Commission des 132 agents de Rusal Friguia 8/12/2012a).

In general, all social benefits given by the company stop with the retirement and the sums emitted by the pension funds are considerably low. The figures that I was able to obtain ranged between 500,000 and 700,000 GNF (70–100 USD in 2015), four times a year (Interview with Journalist 15/01/2014). This regularly subjects entire households to misery: in many cases, the retirees have neither constructed a new house nor saved money and they face considerable problems in abruptly stopping expenditure on their cars and extended families (Interview with Worker CBG 11/02/2014). Consequently, kids often have to leave school and/or migrate (Interview with Activists 15/03/2014), medical bills can no longer be paid (Diallo 2012d), and the families continue to occupy the houses that have already been handed over to other employees (cf. Barry 2008: 37).

In the following years, Rusal and the 132 met several times in court. While the 132 insisted on their rights, Rusal accused them of blocking the rail in September 2009 and squatting in their former houses.[51] It seems that this issue also contributed considerably to the dogged situation after 4 April 2012 (Diallo 2014s, ab). Similar cases at CBK and CBG seem to have been resolved by the end of 2012 when the *Friakas* had already been locked out.[52]

Wages

The existing literature and several interviewees mark the start of the downfall of "Little Paris" already in the mid-1970s, with the crackdown of the regime Sékou Touré on the so-called counter revolutionaries of Fria. According to a Guinean newspaper publisher, purchasing power in Fria has slumped again considerably since the 1990s (Interview with Journalist 15/01/2014; cf. Storozhenko 2009). Together with about a dozen of long-serving workers, I was able to substantiate this decline based upon individual wage bills that we compared with their value in dollars, CFAs, and their purchasing power in terms of rice (Interview with Journalist 15/01/2014). A complete set of wage bills (altogether 223) of a former electrician at Friguia from 1993 until 2012 shows a similar picture (Interview with Engineer Rusal 2/04/2014). Despite starting as a simple worker and climbing up the salary ladder

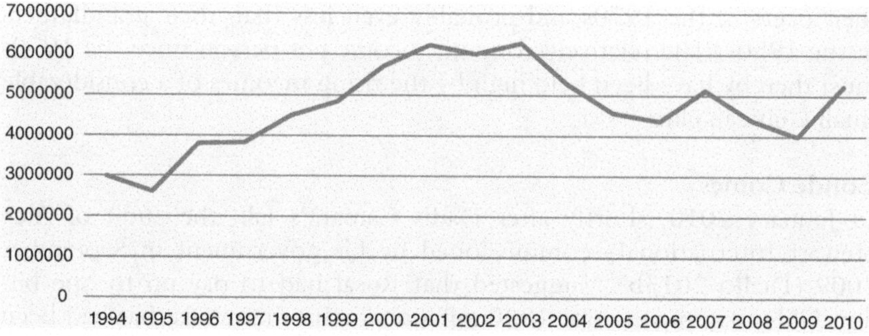

Fig. 3.3 Real net wages of a technical supervisor at Friguia 1994–2010 in GNF (1994 = 1)

continuously to become foremen (category 7; Journalist 17/01/2014) in 1998 and supervisor (*agent de maîtrise*) in 2005, the employee's real annual net remuneration (the sum of his net wages) stagnated from 2000 to 2003 and fell since then (see Fig. 3.3):

The only reliable figures that I found about wages in connection with prices before the 1990s were provided by Sékou Touré before he came to power. According to him, a kilogram of bread cost 55 franc in 1953. The minimum wage stood at 134 francs per day at this time, while a qualified miner at the *Compagnie Minière de Conakry* earned 15,600 franc (Touré 2010: 10, 25, 32, 75). Today, a kilogram of bread costs 12,000 GNF, the minimum wages stand at 440,000 GNF per month (about 20,000 per day), and the basic salary of qualified workers in Friguia stood between 800,000 and 1.1 million in 2010. The national minimum wage thereby stood at about 2.4 kg bread per day in 1953 and fell to 1.6 kg in 2012. A qualified worker earned 116 kg bread per month in 1953, while the basic salary of qualified workers of Friguia was only between 67 and 92 kg of bread in 2010.[53] This would mean that the few paid mining workers in the 1950s already earned more than the miners of Fria in 2010. As Guinean average income doubled from the 1950s until the 1970s, it is also very likely that the wages of the miners of the *Compagnie Minière de Conakry* rose steadily until its closure in 1966. Real GDP per head measured in PPP doubled from 1950 to 1976 and has stagnated since then, with a slight downturn in the 1980s and 1990s (Gapminder 2015). Contrary to this fairly constant GDP per head since the 1970s, even the few current miners seem to earn less than their peers in the 1970s and probably even less than their grandfathers in the 1950s. The relatively constant income per person since the 1970s must thereby have been held high by the rising incomes of a considerably small Guinean elite.

Condé Comes

In January 2010, shortly after Dadis Camara's fall, the audit of Alex Stewart International—commissioned by his government in September 2009 (Diallo 2015b)—suggested that Rusal had to pay up to one billion USD in compensation (830 million worth of minerals that had been mined since 2006, plus up to 170 million for environmental damage) for the losses of the Guinean state due to the privatization of Fria in 2006.[54] The charges of Alex Stewart International for the audit of about 3 million USD had apparently been paid by Chinese investors (CIF) who were competing with Rusal over the stakes in Guinea (Burgis 2010; Diallo 2010: 38; Guinéenews 2015). In June 2010, Deripaska visited Guinea again to talk with the new strong man Sékouba Konaté, de facto president from December 2009 until December 2010, who also seemed to be convinced of the genuine intentions of Rusal in Guinea after the talk and set in place a series of meetings of reconciliation (Diallo 2010: 34). According to a Guinean journalist, another 2-week strike in April 2010 that went along with the hostage-taking of one of the senior officers led to a wage increase of 50% from 2010 to 2011 (Sidibé 2013k; Journalist 19/01/2014; cf. Diallo 2012m).

The coming into power of Alpha Condé in December 2010 did not seem to improve the standing of Rusal in Guinea. Already during his election campaign, Alpha Condé had promised to renationalize Fria (Sidibé 2013a) and after his coming into power he repeated Dadis Camara's claim for up to one billion USD in compensation for the reportedly fraudulent privatization of Fria (Diallo 2014m). In addition, he questioned the validity of Rusal's concession in Dian-Dian and put the company on a list of 41 publicly quoted enterprises that owed taxes to the government (Diallo 2014f; Le Figaro 2011). Rusal claimed the vast bauxite reserves in Dian-Dian on the basis that the Soviet Union had already received them in the 1970s, apparently as debt repayment (Diallo 2014f; Diallo and Kaba 2014; Diallo 2012a). This transfer of claims from the former Soviet Union to the private company Rusal remains questionable and again underlines the special relationship between the Russian state and Rusal (cf. Diallo 2014aa). Neither Rusal nor the Russian state had made use of these reserves since then. As part of an overall revision process of mining contracts (Guinéeinformation 2014), Condé also set the Russians a deadline and threatened to sell Dian-Dian to Chinese investors if Rusal would not start investing. This provoked reactions from the Russian foreign minister and also prompted Deripaska—this time together with the Russian mining minister—to visit Guinea again in April 2011. This time, the meeting between the Guinean head of state and the CEO of Rusal went wrong in the first place. According to newspaper reports, the president became hopping mad (*a piqué une colère noire*; see Guinéenews 2015), because the CEO of Rusal showed a lack of respect and sent him out of the room. The minister followed and both left Conakry on the same day. Some journalists argue that this personal affair was the fundamental reason for Rusal's retreat from Fria (Guinéenews 2015). However, things are neither quite clear in this case, as Condé was accused of having used services of Rusal exactly at the time of this conflict (Diallo 2012m; Soumah 2012).

Lights Out: Rusal Leaves Fria

In the meantime, turf wars between Rusal and the trade unions continued. In December 2011, the trade union lodged another strike warning due to failed negotiations (Deguenon and Traoré 2011h; Camara 2012; Sidibé 2013a). The workers' demands concerned wages (300 USD base salary; Journalist 19/01/2014), the living conditions in the city, the situation of the hospital, and the technical situation of the factory (Interview with Mayor of Fria 18/01/2014). Rusal continued to refer to the high production costs of the factory as compared to the low aluminum price. In December 2011, it reported production costs of 293.15 USD for Aughinish as compared to 382.62 USD in Nikolaiev and 398.48 USD in Friguia. The factory in Fria was the most expensive alumina refinery of Rusal by then (Deguenon and Traoré 2011a: 4; Rusal 2011a). This argument was called into question by a former Guinean director of the factory, Lancei Traoré, who maintained that Rusal had not expanded the capacity of the factory as promised. The company must have known before that the present capacity of the factory was not competitive as it dated from the 1950s (Diallo 2015a).

According to an executive who was present at the negotiations, the workers wanted to discuss the wage issue first and refused to talk about the other points before it was resolved. This halted the negotiations several times (Interview with Journalist 19/01/2014). With the exception of CBK, wages had been the lowest of all industrial mines in Guinea. While CBG increased the workforce and raised salaries, Rusal continued to point to the negative effects of world economic development (Interview with Trade unionist 19/01/2014). Local media additionally reported threats made by a Russian executive during these negotiations, showing pictures of his home town before and after the close-down of a factory, succinctly commenting that this was the future of Fria if unionists were to go on with their demands (Sidibé 2013j). Diallo also mentioned "systematic" power cuts since mid-December and delayed payouts of boni and the threat to cut power and water supply if workers went on strike (Diallo 2012c). After these threats and following a meeting between the unionists, Rusal and the prime minister at the end of December, the unionist took back their strike warning (Sidibé 2013j) and confirmed that they would not strike until June 2012 (albeit not in writing; Sidibé 2013i). In return, Rusal promised not to take any measures that would further reduce the workers' income during this time (Diallo 2013e; Sidibé 2013n; cf. Knierzinger 2016: 149).

At the same time, Rusal also tried to enhance productivity by more strongly controlling the actual working time of its employees (cf. Deguenon and Traoré 2011f: 6, 2011e) and to reduce theft by forcing workers to park their cars outside the factory. About one-third of the core workers possessed private cars (Diallo 2012k). Most of them are now parked in front of the story buildings, covered by a thick dust layer (see Fig. 3.4).

The latter measure led to a serious conflict between the directorate and the trade unions. When confronted with the ban of his own car on 1 February without having been informed beforehand, Mahmady Kourouma—the leader of the trade unions in Fria—saw an infliction of trade union rights and promptly started to prevent any car from entering the factory. This also concerned the director of the factory (Diallo 2014d). It needed the prefect of Fria to find a compromise between the trade unions and the directorate. Rusal then lifted the ban for cars for certain trade union members, but at the same time started a dismissal process against the trade union leaders. Some interviewees think that the threat of dismissal against these eight persons was the main reason for the strike on 4 April 2012 that finally led to the lockout (e.g., Journalist 19/01/2014; Sidibé 2013i; cf. Trade unionist 18/01/2014). While Rusal tried to enforce the dismissal of these workers at the national labor inspectorate, the workers themselves lodged another strike warning.

Fig. 3.4 Dusty cars in Fria, April 2014 (JK)

Rusal subsequently urged the national labor tribunal to ban the strike to prevent "dramatic humanitarian consequences" (Interview with Journalist 19/01/2014) of this strike in Fria. The tribunal responded that it would take some time to take a decision and forbade strikes during this time.[55] On 12 March 2012, Rusal addressed an open letter to the government warning that:

"radical youth of the city of Fria start again subversive movements by distributing flyers. There are calls for violence against the expatriate Russian personnel and the executives of the factory. Certain members of the *collège syndical* of the factory of Friguia, who organize provocative activities among the workers of the factory with the aim to stop production, have united with the movement of the young radicals" (Deguenon and Traoré 2011e; translation by JK; cf. Diallo 2014d).

The stronger control of working time by Rusal was answered by the workers with a "test" of the transport system of Rusal in Fria. On 20 March 2012, by referring to the contract stating that Rusal had to ensure the transportation of the workers to the factory, the trade unions called all workers to take the bus instead of cars, motorcycles, and taxis. This resulted in a de facto strike (Sidibé 2013a). Rusal wanted to dismiss the very trade unionists that had concerted the action and warned that:

"impunity leads to the increase of illegal actions that are at the limit of extremism. Due to the lack of rigor of the *inspection générale* and the *inspection du travail* of the economic zone of Fria, riots could be organized in the near future and malicious individuals could use this situation to create instability in Fria and in the country. [...] The tolerance towards the persons who use their status as trade union leaders to organize impertinent actions and destabilize the situation in town is unacceptable. We count on your understanding and your support, and hope that proactive measures from your side to put away those factors of disorder will be taken" (Deguenon and Traoré 2011e; translation by JK; cf. Diallo 2014d).

When Rusal then abolished certain benefits in kind for its employees in Fria, the trade unions—pointing to an infringement of the agreement of December 2011—successfully mobilized for another strike. Rusal and some workers argue that the decision to strike was taken unilaterally by certain trade unionists and was pushed through by the unionists who were about to lose their job. Others argue that the majority of the workers were behind the unionists. According to them, the strike was even one of the most supported since the takeover of Rusal, since all the three fractions of the trade unions—the *collège syndical* Rusal Friguia, the trade union of EMF, and the representatives of the other subcontractors—agreed for the first time on a strike (Interview with Worker Rusal 3/04/2014; Sidibé 2013n). The decision was taken in public and none of the present workers appealed against it. However, some interviewees argued later that there had been beatings of several workers who did not want to support the course of action of the trade unions (Sidibé 2013a). The decision to strike was followed by a remarkable speech of the mining minister on 28 March 2012 in Fria, beseeching the workers not to go on strike. This speech was recorded by the local radio of Rusal:

"I beg you, it cannot be that this situation prompts you to abandon all the sacrifices that have been made in the last 15 months. I beg you, don't do that, we will work out solutions. We will find solutions, please! If I tell the president that you say you will strike, that won't work out. He trusts you [*heckling by workers*], he has already expressed his gratitude [*honking*]. Every time he delivers a speech, virtually all the previous months, he never ends without talking of the trade unions. Thus, let us follow him and we will tackle the real problem, which consists in improving your living conditions. If you want I will invite the trade unions, we will be with the prime minister and will approach the president [*turmoil in the audience, many workers shout deprecatingly*]. No, I said, no, listen! [*shouts*] This is not to negotiate, this is to find a platform to work on it, to move forward concerning your issue [*again disapproving heckling*]. But why? We have to agree, we, we are all, we are all there to defend the Guineans! But if we don't speak the same language, if we don't understand you, how should we defend you? We have to talk, you have to tell us everything! That's what I'm talking about" (Sidibé 2013a; translation by JK; cf. Sidibé 2013n).

Despite these considerable efforts of the mining minister and the interdiction by the workers' tribunal, the *Friakas* went on strike in the night of 3–4 April 2012.[56] Ironically, the Russian executives had already fled Fria before the strike started—to prevent hostage-takings as in 2010—and argued later that returning to Fria would be too dangerous due to threats by the workers (Interview with Journalist 19/01/2014). In one of his blogs, Ibrahima Diallo remarks that before the arrival of Rusal, foreign workers never had to fear for their security in Fria (Diallo 2012h). A first attempt of negotiations to end the strike already started on 4 April, but was not successful. A second attempt was made on 10 April when the two parties met each other, but could not reach any conclusions. On the same day, the CNTG conferred with the president and called for another meeting on 11 April with Alpha Condé, the Russian ambassador, the directorate of Rusal, some members of the Guinean executives of Fria, and representatives of the trade unions. This meeting apparently failed, because the directorate of Rusal refused to sit in a room with the very trade union leaders that it wanted to fire after the blockage of the factory. Attempts by the Russian ambassador to convince the directorate to come inside did not succeed.[57] According to the *Voix de Fria*, Condé became "hopping mad" again and could only be calmed down by "a call from Moscow" (Sidibé 2013b). The radio commentator continued to state that since this incidence, these were no longer negotiations between the workers and the directorate, but rather between two nations (ibid). Ibrahima Diallo also mentioned the intervention of the Russian president (equivalent to the "call from Moscow" in Sidibé's version) to calm down Condé (Diallo 2013e). However, he also portrayed the Guinean president as a mediator who calmed down the crowd outside, who were trying to bring the directorate to the negotiation table by force (Diallo 2012n; Interview with Activist 3/04/2014).

The consequences of the Russian retreat from Fria could hardly have been worse: as Rusal was the only formal employer, the whole city became unemployed from one day to the next. This not only concerned the core workers (1185; Diallo 2012j) and the subcontractors (2000–4000, depending on the definition of employment; Interview with Activist 2/03/2014; Mayor of Fria 18/01/2014), but also the artisans, traders, service providers, and even the peasants around the city. Many peasant families had sent their kids to school with the earnings from rice, cassava, vegetables, and fruits and were now confronted with the choice between a "return" to subsistence farming or migration (Interview with Journalist 15/01/2014; Camara 2012). The NGO ARSYF mentions a total wage bill of 7 billion GNF (about 1 million USD) per month now being missing in the pockets of the *Friakas* (Interviews NGO Reunion 29/03/2014; Activist 3/04/2014; Former executive Rusal 7/04/2014). The number of this heavily concerned population can be estimated at between 60,000—the official size of the urban community in 2014—and 120,000, the estimated size of the prefecture of Fria before the retreat of Rusal.[58] Most of the workers have ten or more children from two, sometimes even three wives.[59] In addition, many of them had *maîtresses* (in the sense of mistresses), took care of additional children from relatives, and assisted relatives outside Fria. The relations of workers and *maîtresses* are socially accepted. In Sangaredi and Kamsar, they meet primarily in clubs on the weekend. This leads to at least 20–30 persons who directly depended on the workers' salaries (Interview with Journalist 15/01/2014). As Fria was the only alumina refinery in Africa, most of these specialized workers did not find any other work (Interview with Engineer Rusal 4/04/2014). Only few of them had accumulated savings (Interview with Administrator Fria 5/04/2014), whereas the rest plunged with the whole city into "total misery" (Interview with Activist 3/04/2014). Commerce fell by about two-thirds and rice consumption per person fell by one-third since the retreat of Rusal (interviews with traders and street hawkers).

Since January 2013, Fria has been without electricity, aside from the market, the two hospitals and three story buildings.[60] This also concerns the whole administration, being responsible not only for the city and the sub-prefecture of Fria, but for a whole prefecture and it had technical reasons: since its construction in the 1950s, the whole city of Fria was alimented by the same generators that also powered the machines of the factory. Already in August 2012, Rusal decided to stop electricity generation during the daytime and since September power cuts moved sequentially from one ward to another (Diallo 2014q, h). The story buildings only receive power, because otherwise, they could not be alimented with water.

The abrupt decline in purchasing power was associated with the further degradation of the most basic services and particularly public health. Already in 2008, the company hospital had no surgeon anymore (Barry 2008) and Rusal had not replaced any of the hospital's equipment since 2007 (Brygo 2009). Good treatment increasingly depended on good personal ties to the corporate management (Barry 2008). This also concerned evacuations in case of emergencies, whose maximum number was fixed beforehand.[61] With the departure of Rusal, workers even had to pay for their medicine and the personnel of the hospital no longer received salaries. As in the case of the guards at the factory, these employees seem to resell drugs and other goods that are still delivered by Rusal since then to make a living (Interview with Activist 3/04/2014; Former executive Rusal 4/04/2014). Before the crisis, only 35% of the hospital's admissions were actually employees and their relatives (Brygo 2009).

This breakdown of the health system went along with a serious degradation of hygienic conditions. As waste was no longer collected, it agglomerated in the city and was burned down, where the people live and work: beside story buildings, hospitals, and administrational buildings (Interview with Journalist 19/01/2014). In an interview with the *directeur général adjoint* (DGA) David Camara on the company radio in May 2013, it became clear that this still was perceived as a part of the responsibilities of the company, whose leaders had left the city more than 1 year before:

"The problem today is a lack of lorries to empty these waste containers. All we can do is to regroup them with our proper means and, unfortunately, the only lorry that was in use had a problem with its gearbox. By the way, only today I urged the mechanics to swiftly finish the reparation. I think the only solution to save us from everything you can imagine is that the prefecture and the mayoralty help us, maybe together with the transport trade unions. They have already made a campaign like this, have sent us a lorry and given us fuel in order to help us to clear everything in the city. Without them, we won't be able to handle this" (Sidibé 2013d; translation by JK).

In an interview in April 2014, a journalist from Fria used this story as an example of the inertness of the public administration. He himself had tried to raise money to rent a truck that would carry away the waste from sensible places like the company hospital rather than burning it there and polluting the air next to the sickbeds: "But they have crossed their arms. They told themselves that the company did everything" (Interview with Journalist 19/01/2014; translation by JK). This refusal could also be related to the fact that this journalist still worked for Rusal when he tried to organize the waste collection.

Another source of danger is tap water, which has previously been consumed directly as drinking water.[62] The workers' flat, where I lived with several relatives of a former employee during my field research, was located in one of the story buildings and was in a desolate condition. The water of several taps came directly out of the walls, because the water taps had broken off and water in the shower continually dropped down from the upper floors. When somebody used the shower above, the water flowed steadily from the ceiling into the shower bath in the floor below and from there most likely to the floor below. The whole stairway of nine floors smelled of vomit and excrement. In January 2014, the lack of maintenance of almost 2 years finally led to several blockages of the drainage system in Fria (Diallo 2014c). The upcoming fear of epidemics was fueled by the start of the Ebola epidemic in the Guinean forest region in December 2013. According to the interviewees, the restricted treatment at the hospital led to numerous deaths, particularly due to HIV. As in most mining towns, the HIV infection rate in Fria is particularly high (Interview with Activist 3/04/2014; Mayor of Fria 18/01/2014; cf. Trade unionist 18/01/2014).[63]

The decline of the company hospital increased the importance of public health care. The budget of the public *hôpital préfectoral* was doubled in 2013 from 518,755,659 GNF (in 2012) to 1,061,567,789 GNF (ca. 140,000 USD), while the number of consultations declined from 10,100 to 7489. This is linked to the decline of the overall population of Fria. Despite this considerable public support, the budget of the company hospital remains considerably higher (Interview with Trade unionist 19/01/2014; Préfet of Fria 19/01/2014).[64] While the number of births in the public hospital has risen significantly compared to the births in the company hospital, serious cases are still forwarded to the latter (Interview withHospital director 1/04/2014; Activist 3/04/2014).

The *hôpital préfectoral* of Fria has 45 beds and employs six doctors (République de Guinée 2011).

According to numerous informants, the crisis also had profound consequences for the of workers' children (e.g., Interview with Trade unionist 18/01/2014; Activist 2/03/2014). According to ARSYF, most of the 9000 workers' children had to be taken out of private schools and are now schooling in the already heavily overloaded public classes, with 100 and more pupils at the primary level.[65] Inquiries at the school administration office did not confirm high numbers of drop-outs (see below). During the first months after the lockout, this perception, nevertheless, even prompted the government to pay 4 months of private school fees for certain workers' families who were considered particularly vulnerable (Interview with Activist 3/04/2014).

Before the crisis, Fria had a relatively effective educational system (Interview with Activists 15/03/2014). In the beginning, the pupils were separated into Guinean and expatriate schools, but quite early, these systems seem to have merged. Larrue reports of the sudden close-down of the "factory school" (*l'école de l'usine*) in 1964 and its replacement by a public school, which prompted many parents with the necessary means to send their kids abroad, apparently mainly to Senegal and Ivory Coast (Larrue 1997: 137). At the end of the 1970s, the *Mission laïque française* reopened a school for expatriates (ibid: 265, 285). With the continuing decline of the number of expatriates in Fria—from more than 500 workers to several dozen at the end of the 1970s—an overlapping structure of private and public schools emerged, with most teachers holding classes or at least giving private lessons for pupils of both public and private schools (Interview with Activists 15/03/2014; cf. Administrator Fria A 30/03/2014). Many new private schools were founded by workers or retirees (cf. Administrator Fria A 30/03/2014; Journalist 15/01/2014). Having themselves been chosen by Pechiney in various parts of the country due to their grades and based upon standardized tests, these workers thereby actively contributed to the common perception of Fria as the city of intellectuals (Interview with Trade unionist 18/01/2014; Engineer Rusal 2/04/2014). As in other Guinean cities, the quality of private schools mainly depends on the level of their "entrance fee" and to a far lesser extent on the capabilities of their teachers. Due to school fees and additional costs, private schools only host about half the number of children per class compared with public schools (Association des Ressortissants et Sympathisants de Fria 2012: 12). Public schools only take teachers with respective diplomas and pay salaries of 900,000–2 million GNF, or about 130–290 USD (Interview with Administrator Fria A 30/03/2014), while private educators earn much less and are chosen at best due to their know-how (Interview with Administrator Fria B 30/03/2014). From this perspective, the lockout could also have contributed to the end of a segregated educational system that favored the kids of workers.

However, my own calculations based upon the available data for the *commune urbaine* Fria between October 2011 and October 2013 show that both public and private enrollment rates declined by about 20% in the primary sector during that period, while the secondary sector remained relatively stable. Private secondary schools even gained slightly more pupils, while public secondary schools lost about 10%. This rather points to the conclusion that workers' children filled a gap in public primary schools that was created by non-worker's children totally dropping out of school or migrating. Private secondary schools—consisting of about 75% of workers' kids (Association des Ressortissants et Sympathisants de Fria 2012: 12)—did not seem to have felt the consequences of the lockout until October 2013 due to the economic and social capital of the concerned parents. At the same time, many non-workers seem to have sent their children to relatives in other cities and/or to work on the street.[66]

The financial consequences of the lockout for the local administration are highly contested. While several senior officers of Friguia confirmed that Rusal has continued to pay local and national taxes afterwards,[67] local administrators, and politicians stated the opposite. According to the treasurer and the *secrétaire général* of the sub-prefecture of Fria, the tax payments of Rusal fell tremendously after 4 April 2012, which led to contracting budgets and considerable debts with building enterprises (Interview with Administrator Fria A 2/04/2014; Administrator Fria B 2/04/2014; cf. Journalist 19/01/2014). Due to relatively high labor costs, the municipality had almost no money left for infrastructure projects. In 2013, the *commune urbaine* paid 65 contract workers, comprising school and hospital wardens, a special municipal police force (see below), and health workers (Interview with Administrator Fria B 2/04/2014). According to the calculations of the journalist Akoumba Diallo, the urban community received 40% of its budget from local taxes of Rusal. The budget of the prefecture—which almost entirely comprises wages—depended to 75% on Rusal's taxes (Diallo 2010: 12).

From Fria to Dian-Dian

Already in May 2012, the dark clouds over Fria seemed to open up. In the local radio station—which was still run by Rusal—the mayor transmitted the promise of Alpha Condé that if the workers of Fria took "up work again", he would "take care of the rest" (Sidibé 2012). Shortly thereafter, the strike leader Mahmady Kourouma appeared on the same radio, also reporting very optimistically on a meeting with the president (Sidibé 2012d). Condé's sureness was due to a similar announcement of the Russian ambassador that Rusal would start paying salaries again, repay its debts to certain subcontractors, and restart work in the factory, within 1 week (Sidibé 2012a). The workers accordingly suspended the strike on 26 April (Diallo 2012f)[68] and the conflict partners resumed negotiations, although unfortunately, no solution was to be found (Sidibé 2012b). While the trade unions insisted upon increased salaries, Rusal changed course again and declared that even if a compromise with the trade unions would be achieved, it would only relaunch the factory if aluminum prices were reasonably high (Sidibé 2012c). Ibrahima Diallo cites an article in the Guinean Lynx (Issue number 1049)—published already on 22 May 2012—which contained the announcement of Deripaska himself that Fria would only restart if the aluminum price would reach again 2400 USD per ton, proceeding to state:

"if Guinea insisted, this would be at the expense of international relations, both between Guinea and the investors, and between Guinea and the international community" (Diallo 2014p; cf. Diallo 2014g; translation by JK).

Thereby, the company made it official that the strike was not the only reason for the shutdown of Fria and even added further conditions to this, notably that (1) the workers assume the illegality of the strike and thereby forego their salaries until June 2012, (2) no further strikes or blockades would take place, (3) wage negotiations would be resumed at the earliest in December 2013, and (4) the security of the Russian expatriates would be guaranteed.[69] Confronted with this turnaround of Rusal, the Condé government was embarrassed and tried to gain back the workers' support by paying three consecutive salaries on its own, from October to December 2012. When confronted with the demands of the government to refund these payments, Rusal stated that it had never agreed to this procedure.[70] On 1 September, the trade unions, nonetheless, accepted the four conditions of Rusal: workers did not claim outstanding salaries until June, guaranteed the security of Russian personal, and promised not to strike anymore and to wait for wage negotiations until December 2013.[71]

In the following months, Rusal augmented the pressure on the Guinean state. While profiting from the situation in Fria—a formal stop of production due to the low aluminum price would have been much more costly—the company also tried to get guarantees for its claims in Dian-Dian. If we follow the records of Ibrahima Diallo, in the late June 2012, Rusal had already announced its plans to shut down the factory, provoking counterthreats from the ministry of mines, suggesting that it would strip the company of its assets in Dian-Dian and in Débélé if it went ahead with these plans (Diallo 2013e). This, in turn, led to the political intervention of the Russian government, which threatened to suspend negotiations on the HIPC Initiative.[72] This threat existed until 26 September 2012 when Guinea was granted 2.1 billion USD in debt relief (International Monetary Fund 2012b).

The population of Fria naturally did not approve this inconstancy of Rusal. The city witnessed several riots with demonstrators erecting and dismantling roadblocks (28 August 2012) and so-called *loubards* (yobs) attacking traders and destroying market stands and shops in search for food and valuables (September 2012). The following clashes with police and gendarmerie caused several deaths (Interview with Activist 3/04/2014; Diallo 2014q). In July, a group of young men blocked the train station of the factory and demanded diesel "in order to feed their families (Diallo 2014q) ." The few workers still being employed by Rusal to ensure the supply of water and electricity had not been paid since the start of the strike. Following threats to stop working by early August (Diallo 2014q; Sidibé 2012e), these workers went on strike in October, leaving the city for 3 days without water (Sidibé 21/10/2012c). This strike went along with demonstrations by the "unemployed" workers who demanded the promised salaries of the government and also tried to stop the provision with water and electricity, thereby provoking the prefect again to increase the police forces of the city (Diallo 2014q; Sidibé 2012f). In December, trade unionists—accompanied by parts of the former executives of Friguia—informed the government that Rusal no longer abided by its social obligations. In view of the extreme poverty in the city—caused by a lack of water, electricity, and public hygiene—the two groups called upon the government again to annul the privatization of 2006 (Diallo 2012l; Balde 2012). However, on 28 December 2012, the company finally had its way: the Guinean government and Rusal signed a mining contract for Dian-Dian, stipulating the construction of a bauxite mine of 3 million tons per year until 2015 and feasibility studies for a further extension of output to 9 million tons and the construction of an alumina refinery. This decision was ratified by the transition government and put into effect by presidential decree in March 2013.[73] The company's promises to relaunch Fria with the resources that had been liberated on the stock markets with the permit of Dian-Dian have not materialized at the time of writing (Diallo 2013b, 2014g).

By the end of 2012, the effects of Rusal's retreat on purchasing power and social services had already reached the bottom of human needs: families had sent their children to relatives or on the streets to work and had cut their expenditure on rice per head by one-third. Interviews with about 50 street hawkers and owners of small boutiques showed that rice consumption was about 300 g per person per day before the crisis—with relatively small variance[74]—before falling to 220 g since 4 April 2012. At the same time, the interviewees' profits fell by two-thirds (Interview with Former executive Rusal 3/04/2014). Many families now only ate once per day rather than twice, as before the crisis.[75] The correlation between income and caloric intake was quite clear: only when one member of these families still had a job—for instance, in the administration, as a teacher or a security guard—had the rice consumption not slumped. The average family size in the survey fell slightly from 13 before the crisis to 11.75 in April 2014. Families with fewer members also consumed less rice per capita. This means that the bare availability of rice seems to have strongly influenced migration: families sent their members to relatives or to other cities and construction sites to earn money. Merchants who traded in rice and other staple foods became rare during the crisis, because they were often forced to share. The debts of workers thereby became debts of these merchants versus the importers. Therefore, in order to maintain their "dignity" and their business, many of these traders stopped the trade in staple foods (Interview with Salesman 6/04/2014; Sidibé 2013h). The continuous donations of rice since then provide another strong indicator that Fria de facto experienced a hunger crisis after the retreat of Rusal.[76]

On 11 November 2012, recurrent reports on the theft of fuel prompted Seydouba Camara—prosecutor and *juge de paix* (justice of the peace) of Fria—to start investigations. When one of the alleged thieves had been seized, Camara started investigations himself, reportedly due to a lack of confidence in his own security forces (Diallo 2014q). Together with several gendarmes, he was guided to a store room, where the fuel had been collected. According to several reports, he then parked his car and ordered the owners to open up the store room in the meantime. When he came back, his auxiliaries were gone, their motorbikes stood in flames, and young men around the store room started to attack him. They destroyed his car, threw stones on him, and pursued him throughout the city. The judge only escaped by jumping over the walls of one compound to the next. Furthermore, a young journalist who tried to take photos of the burned vehicles was beaten and his own car was set on fire (Interview with Journalist 19/01/2014). On the next day, the confiscated fuel had disappeared again and the judge's house had been ravaged, also leaving his documentation and evidence destroyed (Diallo 2013b; Sidibé 2012i; Journalist 19/01/2014).

Only few people had obstructed this ambush of 11 November. This can be related to the fact that high military officials—including the *conseiller militaire* of Fria—were perceived as being implicated. Several sources also suggest that the majority of the *Friakas* thought that these thefts were justified or at least tolerable. In October 2012, the *directeur général adjoint* (DGA) David Camara talked openly on the local radio about the discrepancy between the fuel sent by Rusal and the actual use for electricity generation. While the executives in Fria continued to argue that this related to the old machines, everybody seemed to know that there was a parallel fuel market in Fria.[77] It appears that this was not a consequence of the crisis: fuel seemed to have been part of constant dubious deals when the factory was still operative. According to trustworthy information of former employees at Friguia, about 70% of all the expenditure of Rusal in Fria was spent on fuel. Concerning the use of this fuel, these employees talked about a "total opacity" (Interview with Former executive Rusal 4/04/2014). Similar fraud on a similar scale is decried behind closed doors in Kamsar (Interview with Entrepreneur 19/02/2014). In addition, the import of petrol also seems to aliment the pockets of government officials in Conakry. Already in 2009, Rusal argued that it could reduce its costs of fuel by 18% if it was allowed to import fuel on its own rather than having to buy it from the government

(Storozhenko 2009) . During the first half of 2013, several other cases of theft were reported, mostly concerning the group of about 120 employees[78] who still worked for Rusal as guards and technicians to guarantee the supply of water and electricity. As these men had not been paid by Rusal, they threatened several times to abandon their work (Diallo 2014h, r) and at least some of them seemed to take their reward in kind in the form of fuel, sodium hydroxide, cables, etc. On several occasions, arrested employees had been freed again shortly after their detention due to threats against the police forces (Diallo 2014r). An audit of the factory in May 2013 finally showed that the whole factory had systematically been plundered during the 12 months since Rusal had left. The DGA David Camara, a Guinean, stated on the local radio that "the factory has not been robbed but looted":[79] The "thieves" seemed to have removed anything that was removable, including cables (e.g., 2 km of cables of the crusher), lubricating oil, and all sorts of equipment. This also meant that electricity could only be provided again to the city after reinstalling the cables and the power transformers that had been removed.[80] At the end of May 2013, Rusal seemed to regain control over the security forces: after having compiled a list of 59 suspects reportedly involved in the theft of fuel, the police detained 23 guards who had been working for Rusal at the factory (Diallo 2014r) .

When the first anniversary of the lockout approached, the mood in the city seemed to change again. After three salaried months from October to December 2012, the workers were again left with nothing, causing an economic crash of the whole city again (cf. Sidibé 2013). Continuing rumors and false promises of a restart of the factory worsened the atmosphere. In January 2013, the Guinean DGA announced a restart of the factory for April or May (Diallo 11/08/2014q). However, when the Guinean mining minister repeated the promise to restart in March, representatives of the company denied this again on the local radio and stated that "the minister is not master of the situation" (Sidibé 2013). On 4 April 2013, workers and their partners marched peacefully to the factory and then organized a sit-in in front of the prefectural administration, demanding their salaries and singing "the prefect is a liar!" (Diallo 2014r). Already on the first day, the place was renamed "Place Tahrir". The crowd prevented the whole administration from entering the buildings—the local radio even talked about the administration being held hostage (Sidibé 2013c)—and urged the DGA to cut electricity for the hospitals and the story buildings. The DGA obeyed the demonstrators and thereby triggered heated discussions about the pertinence of this measure. Condé initially reacted by deploying more security forces and then also sent half of his cabinet to calm down the demonstrators (Diallo 2013h). Despite discussions with the ministers and various interventions of religious leaders, elders, and NGOs, the sit-in continued until 14 April 2013. On 12 April, a delegation of the Russian government arrived in Conakry and negotiated with the Guinean government. This was followed by negotiations with Rusal, which then decided to start with an audit of the factory within the next 2 weeks. These guarantees apparently prompted the demonstrators to end the blockade on 14 April. Neither men nor property had been harmed during the occupation.[81]

How to Jump Off a Lion

However, despite all these efforts, the factory had not been relaunched at the time of writing. With substantial backing by the Russian government, Rusal seemed to have gained the upper hand both in conflicts with the *Friakas* and with the Guinean government. The president's efforts to relaunch the factory appeared increasingly futile and called into question the authority of the government in general (Interview with Police 28/03/2014). Many believed that Condé was not apt to confront a transnational company like Rusal that was heavily backed by Russia, all the more as the government was wrestling with other mining giants like Rio Tinto at the same time (e.g., Journalist 15/01/2014). It did not take long for the government to buy into this discourse. In September 2013, Prime Minister Mohamed Said Fofana declared that "we can do nothing against Rusal, the owner of the factory, it is more powerful than the state" (Diallo 2013m; translation by JK).[82] Half a year later, Fofana also compared the situation with the Guinean metaphor of a man sitting on a lion and being admired by the crowd, while only thinking of how to jump off the beast without being devoured. Ibrahima Diallo commented on this metaphor:

"But the question is: who is the cavalier? And who is the lion? There are two answers. The first one is that the prime minister is the cavalier and the head of state is the lion. The second one is that the cavalier is the head of state and the lion is Rusal. Which is closer to reality, because in view of what is going on in Fria, the master of the game is clearly Rusal, which acts as it sees fit in total impunity" (Diallo 2014x; translation by JK).

Ultimately, it seemed that the government deemed future investments of Rusal and the relations with Russia more important than the (modest) loss of tax income and the political danger of an uprising in Fria (Interview with Former executive Rusal 3/04/2014). This relation between a seemingly almighty company and a powerless state is even more intriguing if compared with the medial (self-)portrayal of Putin's relation with Rusal. In one of his blogs, Ibrahima Talibé mentions a YouTube video of 2009 showing Putin as the savior of the Russian town Pikalevo, whose population was in a situation quite comparable with Fria. After heated protests due to the close-down of an alumina refinery and two cement works in Pikalevo, Prime Minister Putin flew in and met the owners in front of the camera, including Oleg Deripaska. After calling them greedy, unprofessional cockroaches, he forced Deripaska to sign an agreement that stipulated the restart of the factory. Deripaska had to sign the agreement by bending down in front of the sitting Putin and was subsequently even called back by the president, because he had unintendedly taken his pencil. Particularly, Putin's last nonchalant words "give me back my pen"—resembling the infamous words of Arnold Schwarzenegger in the action movie Terminator—seem to have made a significant impression in Fria. In his comment on the video, Diallo did not mention the possibility that this video could, in fact, have been faked with the consent of Deripaska, but rather portrayed Putin as a role model for Guinea (Conakry Infos 2014; Diallo 2014r; Sabaté 2009). One cannot help but to compare this awkwardly cool Russian president with the embarrassing behavior of Captain Moussa Dadis Camara in a very similar situation in his own TV show. Being confronted with the neglect of the factory in Fria by Rusal, he invited its representative in Guinea, Anatoly Patchenko, only to publicly shout him down in a most violent manner. After several minutes of meaningless shouting followed by periods of silence, Dadis finally demanded that Patchenko should promise that Rusal would not close down the factory. Patchenko indeed promised, only to leave the country shortly thereafter (cf. Saada 2009).

On 30 April 2013, Guinean courts declared the lockout by Rusal legal and confirmed the illegality of the strike (Cherie FM Guinée 2013; Diallo 2013m), which meant that Rusal did not have to pay salaries. Thereupon, the company promised again to restart the production in Fria (Camara 2013) and even remitted an allowance of about one salary (2.9 million GNF) for every worker as a "gesture of good will". Provided that they signed a statement declaring their responsibility for the shutdown of the factory and express their apologies, Rusal even promised to pay more. In order to be eligible, the workers had to sign the following statement:

"We, the workers of RUSAL/FRIGUIA, acknowledging what has happened, are aware of the fact that our indifference and our wait-and-see attitude have contributed to protract the crisis despite some good will, especially shown by the *commission de l'encadrement*, female workers, women's associations, people of faith and the regional coordination of FRIA. For this reason, we ask our employer for TOLERANCE AND PARDON for all the misconduct that happened during the crisis. Therefore, we, the workers of RUSAL/FRIGUIA will strictly obey internal regulations, the general rules of discipline within the enterprise and contribute to maintain a climate of peace and calm in all respects within the enterprise and in the city" (Le Guinéen 2013; capital letters in the original, translation by JK).

In a broadcast of the *Voix de Fria* 2 weeks before this campaign, the DGA David Camara again conveyed threats of Rusal if this statement would not be signed by the majority of the workers:

"Otherwise, they will pack their bags and leave, because the demand for alumina is strongly decreasing, with consequences on the aluminum sector, where Rusal has to reduce its production by 300,000 tons, which is the equivalent of our annual production of 600,000 tons of alumina. This shut down would therefore be in no way a loss for them" (Transcribed and translated from Soussou into French by Diallo 2014aa; translated into English by JK).

Already before this petition, about 80% of the workers had sent their apologies to Rusal and distanced themselves from the trade unions (Camara 2013; Sidibé 2013e). The petition itself was then signed by over 90% of the workers, although some of them were, nonetheless, excluded from assistance payments. The DGA David Camara explained the reasons for this exclusion on the local radio:

"Rusal told me that it's their house, their property. They will give their money to whom they want. Trouble-makers and those who have positioned themselves wrongly towards Rusal won't get anything" (Sidibé 2013f; translation by JK; cf. Diallo 2014g) .

Those who had refused to sign the statement and still wanted the assistance payment had to answer a questionnaire about their motives, which would determine their eligibility (Diallo 2014aa, b). David Camara also promised again on behalf of Rusal that the factory would start producing again. By the time the necessary modernization measures would start, the involved workers would receive their full salaries again, while the rest would receive 50% (Sidibé 2013e). At the time of writing, Rusal had neither started the modernization nor remunerated the workers as promised (Diallo 2013j, 2014h, g).

The Women Take Over

After the Guinean court decisions, the only legal resort left for the workers in case of further unfulfilled promises was the claim that Rusal had never defined their status: they were neither employed nor unemployed, nor were they put on leave. The term used by Rusal, *chômage technique*, did not exist in Guinean law (Sidibé 2013g; cf. Interview with Activist 3/04/2014; Diallo 2014aa; Trade union leader 3/03/2014). In the following months, despite being deprived of both their salaries and most basic services, the workers' movement was caught up in disputes concerning the next steps to take (Diallo 2014g). This void was filled by the workers' wives and female workers. When both power and water supplies were cut for six consecutive days in November 2013 (Diallo 2014t), the women protested again on the "Tahrir place" and secured the promise of the government to construct several well drillings (Diallo 2014g, t).

On 2 and 3 December 2013, the workers' wives and female workers blocked the streets of Fria again and demanded the government to clarify the status of the workers. On the night of 3 December, the barricades were violently destroyed and the women were attacked by Guinean citizens, apparently predominantly traders. The women then marched to the prefecture, where they were harassed, beaten, and dispersed with tear gas by the gendarmerie (Diallo 2013m, 2014t; Sidibé 2013l). In a following radio broadcast, a worker summarized the demands of the protesters as follows:

"The movement will continue, because, as I have explained, we have nothing to eat, we don't know our administrative status. I have done 34 years of service, should I leave like that? Without any word, without any remuneration, without compensation? At least I have the right to receive my pension, I have rights if I am fired, I have the right to an indemnity. Because I have stolen nothing! What I did was constitutional, I was on strike." (Sidibé 2013l; translation by JK).

In January 2014, Rusal finally began to pay a small assistance to the employees of Friguia and of EMF, apart from around 50 "troublemakers" (Interview with Activist 3/04/2014; Mechanic 4/04/2014).

The workers received 1,050,000 GNF a month, which corresponded to about 150 USD and was far below the sums promised in April 2013 (Interview with Former executive Rusal 4/04/2014; Diallo 2014d; Journalist 25/03/2014). These payments began after another round of negotiations and the signing of a contract between Rusal, the Guinean president, national trade union leaders, and representatives of workers and executives of Friguia.[83] In exchange for (renewed) guarantees to pay 50% of the salaries already before the restart of the factory, Rusal was granted the use of the rails of CBG for its new mines in Dian-Dian. As a large part of the factory was deemed beyond repair, the company additionally promised to immediately start a feasibility study for the restart of the factory with coal (Guinéeinformation 2014; Diallo 2014ab). These negotiations probably took place, because Alpha Condé was about to name a new government after having won the legislative elections in September 2013. The mining minister Mohamed Lamine Fofana was replaced by Kerfalla Yansané, who had been financial minister since the transition government under Jean-Marie Doré and who again called into question the privatization of Fria and the concession of Rusal in Dian-Dian (Interview with Journalist 25/03/2014; cf. Diallo 2014c). The payment of this small "assistance financière exceptionelle" again depended on a statement that had to be signed by the receiver:

"Hereby I ask you to grant me an exceptional financial assistance, because I am in a very difficult financial situation. Through this request, I pledge to refrain from taking part in any illegal actions and all other disloyal activities against the company FRIGUIA S.A, and above all those that could endanger the security of foreign specialists and their property. I personally pledge, as far as possible, to contribute to the maintenance of a climate of peace and calm in every respect within the factory and in the city of Fria" (Diallo 2014r; translation by JK).

In the meantime, another riot took place in Fria. After Tafsir Sylla, an unemployed youth, was beaten to death by the *police municipale*, clashes between the forces of order and the "youth" led to the burning of the buildings of police and gendarmerie, as well as lootings of administrational buildings of the municipality and the prefecture (Diallo 2014x; Nouvelles de Guinée 2014: 5). The mob first began to throw stones at the buildings of the *commune urbaine* and then attacked the headquarters of the gendarmerie, freeing several detainees, and then lighting up the building. From there, the protesters marched to the police building and the civilian prison to do the same (Diallo 2014x). The lootings in these and other public buildings led to the theft of 30 firearms, 24 diesel generators, several bottles of strychnine from the agriculture department, computers, printers, etc. On the next day, the police struck back and searched houses for the goods and weapons, while the government announced another delegation and once again promised to do what it could to restart the factory (Diallo 2014x; Administrator Fria B 30/03/2014).

The *police municipale* is a paramilitary group that was formed by the municipal government and mostly comprised young disadvantaged men who had barely been trained for this task and were engaged in gang conflicts themselves before entering the group. According to a member of this police force, the only training that they received before starting to patrol the city was "karate lessons" (Interview with Member of the municipal police 1/04/2014). The interviewed member of the *police municipale* had not been paid since the start of operations in early 2013. Tafsir Sylla is said to have been the member of a rivaling gang that had not been accepted into the *police municipale*. The beating took place after the victim had been caught consuming "drugs". The drug itself had never been specified officially, but the interviewed member of the police force stated that Tafsir Sylla had been beaten to death for smoking marihuana (Interview with Member of the municipal police 1/04/2014; cf. Nouvelles de Guinée 2014: 5; Diallo 2014x). This incidence was the sixth since the retreat of Rusal, which prompted the government to send military troops to Fria (Interview with Journalist 3/04/2014; Journalist 3/04/2014; Mayor of Fria 18/01/2014).

Current Situation and Outlook

Since this incident, the situation in Fria remained relatively calm. In July 2014, Rusal announced the start of constructions in Dian-Dian.[84] Rather than clarifying the status of the workers, settling its debts with its subcontractors, restarting to supply the city with water and electricity, paying for social services, and reacting to the numerous local demands for compensation, Rusal just continued to pay about 150 USD per worker as voluntary assistance, as well as providing electricity and water to the story buildings, the hospital, and the market place (cf. Diallo and Kaba 2014; Diallo 2014z; Rusal 2014c). Having waived on all salaries that had been lost during the strike and having received four basic salaries until the end of 2012, the workers had not been paid for more than 1 year (the whole year of 2013 and 2 months in 2012) and are now temporarily tranquilized with a small allowance (Interview with Journalist 15/01/2014).

The *Friakas* have, nonetheless, retained a spirit of hope. With the financial assistance of *Friakas* in Europe, the US, Canada, Russia, Angola, and Morocco (Interviews NGO Reunion 29/03/2014), many projects have been started since the lockout, most of them envisioning a "future of Fria without mines" (Interview with Activists 15/03/2014) that relies on agriculture, tourism, and small manufacturing.[85] These contacts and a markedly intellectual workers' culture in Fria distinguish this city from other Guinean agglomerations. During my spare time in Sangaredi, another bauxite mining town in Guinea, I tried on several occasions to find something to read, but found neither newspapers nor novels. The only books to buy directly in the markets were school books. On the contrary, one of the first interviewees I encountered in Fria was a former worker who had opened up a bookshop (Interview with Engineer 18/01/2014). A few hours later, a blue-collar colleague of his introduced me to his young daughter, who had just won the national spelling contest (Interview with Trade unionist 19/01/2014). In 2014, the NGO ARSYF finished the construction of a small library and was trying to establish a radio "that is private: belonging to the civil society" (Interview with Activists 15/03/2014; translation by JK).[86] The latest projects to draw attention to the situation in Fria were the organization of a peaceful march (Interview with Activists 15/03/2014; NGO Reunion 29/03/2014) and the intention to file a law suit against Rusal with the help of the Guinean minister for human rights (Interview with Journalist 25/03/2014).

Until the lockout, *Friakas* often encountered situations that could be described as consequences of a local paradox of the plenty. The government provided public funds almost exclusively for salaries of administrators in Fria, whereas everything else—such as infrastructure and the equipment of local authorities—was delegated to the companies. This led to the paradox situation, whereby local authorities in Fria were even less well-equipped (for instance, with motorbikes or with computers) and less trained than their colleagues in neighboring prefectures. Because people lived comparatively well from mining, there was neither an urgent need to diversify the local economy—for instance, with agricultural projects—nor with better connections to other trading centers, such as Télimélé (Interview with Trade unionist 19/01/2014; Préfet of

Fria 19/01/2014; Mayor of Fria 18/01/2014). When Rusal left, this resource dependence became apparent.

The first step to end this dependence would be the connection to the national power grid and the construction of a pipe water system independent of the factory (Interview with Journalist 19/01/2014). According to the mayor of Fria, the government has again launched a study in this respect (Interview with Mayor of Fria 18/01/2014; cf. Diallo 2014w). Several interviewees still believe that Rusal will come back to rebuild the factory and that this would probably take place in connection with the erection of a coal plant.[87] Since parts of the infrastructure already exist and the reserves are far from being depleted, it is most likely that bauxite mining will restart in Fria. Besides a Chinese project in Wawaya in the prefecture of Fria (Interview with Administrator Fria A 2/04/2014), media often report on plans of the International Mining Company (IMD)—a mining company mounted by the former Guinean football player Seydouba Bangoura in cooperation with the Canadian mining company HATCH and the Dutch firm Royal Haskoning DHV (Diallo 2010: 8)—to restart bauxite mining in Fria (Diallo 2013n).

3.1.2 Débélé

The fully state-owned mine of CBK (*Compagnie des Bauxites de Kindia*) is situated in Débélé, a 50 km drive from the third largest city of Guinea, Kindia (Fig. 3.5). It is not comparable to the other mines due to its proximity to this large city, as well as the low living standards of its core workforce. Débélé counts no more than 2000 inhabitants and is administrated as a "cité autonome" in the sub-prefecture of Mambia, which counts altogether 26,500 inhabitants (République de Guinée 2014: 11; Sous-Préfet of Mambia 18/03/2014; Administrators Mambia 18/03/2014). In recent years, the mine was extended to the sub-prefecture of Friguiagbé, which counts 34,500 inhabitants and lies closer to the city of Kindia with 169,000 inhabitants (République de Guinée 2014; Administrator Kindia 24/03/2014; Mayor of Kindia 17/03/2014).[88] Neither Mambia nor Kindia are dependent on electricity or water supply by the mining company (Interview with Activist 17/03/2014).

This leads to considerable autonomy of the political administration versus the mine, which is envied by the political elites of the other

Fig. 3.5 Old mining pit in Débélé, April 2014 (JK)

company towns. Mamadou Conté, the prefect of Fria, insists that "today, the company has no impact on the social life in Kindia. No impact at all! Whether they shut down the enterprise or they open it, this has no impact on the population of Kindia. But here, the opposite is the case!" (Interview with Préfet of Fria 19/01/2014; translation by JK). Besides the proportions, this also relates to the economic alternatives in Kindia and the fact that the mining town itself neither became the capital of the prefecture nor of the sub-prefecture (Interview with Préfet of Fria 19/01/2014). However, Kindia's economic development since independence has neither been very positive. Due to its fertile land, the area had been used extensively for cash crops—particularly bananas—during colonialism, but the separation from France led to a Western boycott of Guinean agricultural products. In 1958, this even resulted in the abandonment of a railroad between Conakry and Kankan, whose remnants can still be visited in Kindia (Interview with Former executive Rusal 3/04/2014).

In the last 10 years, the mining industry has become increasingly important. Besides Rusal, a number of other bauxite companies hold exploration rights in the prefecture of Kindia. By December 2012, the mining ministry listed Alufer, a company registered on Guernsey that also possesses assets in Labé and Bel Air; Anglo African Minerals (through its subsidiary Forward African Resources), registered in Ireland; the Kakandé Alumina Company (through its subsidiary International Mining Development), registered on the Virgin Islands; and a Chinese company cited as "SARL de la zone inter. De coorp. Econ. Et com. Du Quinghai" (République de Guinée 2013d). The Chinese company could, in fact, belong to Chinalco, which announced in 2005 that it had obtained 21 exploration permits, covering 1488 km^2, most of them in Kindia.

According to several interviewees, CBK was both a means for the Soviet Union for debt collection similar to the current barter agreements of China with African countries, and it served as an important asset in the Cold War. More than any other country, Guinea was seen as a bridgehead to Africa, and the large Western bauxite mine of the *Compagnie des Bauxites de Guinée* was already under way when negotiations started between the Soviet Union and Guinea. The exact amount of Guinean debts versus the Soviet Union was unclear when Touré died in 1984 and it remains unclear to date.[89] Jörg Dux argues that this was primarily due to the devaluation of the Russian currency and subsequent disagreements after the collapse of the Soviet Union. According to him, the debt had, nonetheless, been renegotiated in 1994 (Dux 2000). The Soviet Union had financed educational programs, built roads, a university and a stadium, sent teachers, scholars and doctors, and pushed forward electrification. All of this was to be repaid with Guinean bauxite, for which the Soviet Union built a refinery in Nikolaiev (Ukraine), which began production in 1980 and continues to operate today. This factory was technically geared to Guinean bauxites, and was partly constructed by Pechiney, which then also bought considerable quantities of aluminum from this and other Russian smelters (Pechiney Ugine Kuhlmann 1976: 7, 1980: 3). In 2008, 65% of the production of Débélé went to Nikolaiev (International Monetary Fund 2008: 7). From there, the aluminum atoms from Débélé mostly migrated in the form of Al^2O^3 (alumina) to the smelters of Rusal on Russian territory. One of these smelters was (and remains) situated in Sajanogorsk, near the Russian borders to Kazakhstan and Mongolia, where the alumina is processed into foil for food and pharmaceutical packaging, as well as direct household and technical use (Diallo 2010: 36; Rusal 2014f). Various actors mentioned that the substitution of the Guinean bauxite would mean costly adaptions to the factory of Nikolaiev, but this does not mean that the factory is in fact dependent on Débélé (Interview with Former executive Rusal 25/03/2014). During a production stop in Débélé in 2011, the bauxite was replaced by shipments from Porto Trombetas (Brazil) and Kamsar (Soloviev 2011: 2), and in a strategy paper in 2013, Rusal also proposed a replacement of Guinean bauxite in Nikolaiev by Guyanian bauxite (Interview with Trade unionist 25/03/2014). However, in 2015, the Ukraine nationalized the refinery in reaction to Russia's annexation of Crimea (The Moscow Times 2015). Due to the technical gearing of the sites, this also endangered Rusal's activities in Guinea and necessitated considerable downstream adjustments. The factory was, nonetheless, outdated. Production costs in Aughinish, another refinery in possession of Rusal in Ireland, are about 25% lower than in Nikolaiev.[90]

The initial agreement between Russia and Guinea concerning CBK aimed at the production of 2.5–3 million tons of bauxite per year and included 100 km of railways across the Kakoulima chain, the renovation of port platforms of the former mining company in Conakry, and the construction of a town and a depot in Simbaya in exchange for access to mineral resources. For this, a parastatal comparable with the MARG of Fria was founded, which was to be operated by Guineans with technical assistance by Russians (Soumah 2008: 152–153). The mine started production under the title *Office des Bauxites de Kindia* (OBK) in the early 1970s. The exact year of the start varies in the literature between 1972 and 1975.[91] The Soviet Union received 50% of its output directly as debt service and 40% was to be sold to the Soviet Union based upon a long-term clearing agreement, i.e., to buy goods destined to OBK. Most of the rest was also sold to other Eastern European producers, mainly due to the integrated structure of the Western firms, as well as due to lower quality. Campbell tried to ascertain whether Guinea profited from this barter trade, reaching a rather negative conclusion. Based upon the available figures, the goods sent by the USSR in exchange for the bauxite were less valuable than the Western price for bauxite.[92] Soumah remarked that the "equipment put on site was powerful and quite competitive", but numerous accidents, nonetheless, led to the nickname "Obkana" instead of OBK, which means "destroy everything here" in Soussou (Soumah 2008: 152–153). In the second half of the 1970s,

OKB lost seven locomotives and 97 wagons due to accidents. Additional problems with logistics and power supply led to considerable losses until 1982 (Dux 2000).

The dissolution of the Soviet Union subsequently led to a decade of relatively obscure relations of control. Still in 1990, the USSR planned to undertake a large development program with the aim of erecting an alumina refinery and an aluminum smelter. The bauxite for this complex was to be mined in Dian-Dian near Boké. However, due to the dissolution of the Soviet Union, production instead fell considerably in 1991 and only briefly recovered when the state obtained a contract with the newly formed Ukraine. From 1993 to 1994, production halved again from 2.3 million to 1.2 million tons of bauxite because of technical problems. Due to the low quality and the low ore grade of the bauxite, the cost of production was higher than the price obtained on the international market. OBK was transformed into a limited liability company and renamed *Société des bauxites de Kindia* (SBK), but a search for new investors was not successful.[93] It seems that Russian actors took over control again in the late 1990s. Already in 1998, the trade unions were negotiating with Russian managers, during a time when the ownership of Soviet aluminum assets was still highly contested. Rusal was only formed 3 years later, in 2001. The Russian delegates, nonetheless, already had concrete plans to reduce salaries, which were already the lowest of the sector, as well as replacing old Guinean executives with young Russians. This led to a 27-day strike in 2000 and the dismissal and 6-week imprisonment of seven trade unionists (Interview with Trade union leader 3/03/2014; Diallo 2014g).[94] Among these prisoners was Mamadou Mansaré, the current *secrétaire général adjoint* of the CNTG, the largest Guinean trade union, who had started working as an engineer at OBK in 1983 and soon became trade union leader at the firm level.

Mansaré connects the takeover of Rusal in 2001 with the occupation of several southeastern Guinean prefectures by rebels from Sierra Leone and Liberia. According to him, these attacks took place only shortly after Rusal had announced its interests in SBK. Conté's army was in bad shape because the president feared for his disposal. He had only sustained good equipment for his personal "praetorian guard", the same that was to kill more than 150 demonstrators in November 2007. During this dangerous situation, a certain Anatoly Patchenko appeared on the scene—at this time a translator at SBK—and apparently proposed to repair old Russian tanks and helicopters of the Guinean army, as well as sending four new helicopters and ammunition. According to Mansaré, he demanded control over SBK in exchange for this help and both sides kept their promises. Accidentally or otherwise, one of the suppliers of the rebellion seemed to be a Ukrainian, who was arrested in Russia at the time when the lease of the facilities was signed (Interview with Trade union leader 3/03/2014).

Rusal took over the management of the company under the name *Compagnie des Bauxites de Kindia* (CBK) and Anatoly Patchenko remained in Guinea until 2009 when he was invited to Dadis's show and left the country shortly thereafter (République de Guinée 2013c: 7; Rusal 2014a). The takeover did not proceed with a privatization of the company (Interview with Trade unionist 25/03/2014). Rusal only rented the facilities, although given the unclear previous situation the workers seemed to be pleased that there was a new investor (Interview with Executive CBK 21/03/2014; Employee CBK 18/03/2014). However, when tensions increased between the government and the company, Rusal was accused of only paying insignificant amounts for the infrastructure that it used. Officially, the company rents the facilities in Débélé, the whole railway to Conakry, the port and the offices, workers' homes, and other infrastructure in Débélé and Conakry (République de Guinée 2006: 9). According to the former Prime Minister Lansana Kouyaté, all of this amounts to no more than the rent for a "simple villa" (Diallo 2014f; cf. Interview with Activist 14/03/2014). In addition, Rusal had obtained a comprehensive tax exemption until 2003 (Interview with Executive CBK 20/03/2014).

Today, Rusal mines about one-third of its bauxite in Débélé (Rusal 2014a). The production price ranges between 10 and 15 USD per ton and is thereby much lower than the one of CBG (20–25 USD/t; International Monetary Fund 2008: 9).

3.1.2.1 Tax Payments and the Electrification of Mambia

Only a few years after the end of the tax holidays in 2003, Rusal stopped paying local taxes again. The explications for this refusal, its consequences, and the exact timeline differ considerably depending on the sources. Algassimou Diallo, *directeur adjoint* of the mine in Débélé, explained that Rusal had paid 200,000 USD in local taxes in 2005/2006, which were meant for the construction of a secondary school and a health center. This construction should have been managed by the *commune rurale* of Mambia, but was claimed by the governor and the prefect and did not lead to the constructions (Interview with Executive CBK 20/03/2014). A former civil servant at Mambia also mentioned an electrification project of the Guinean government in 2003 in Mambia that did not work out (Interview with Former Administrator Mambia 24/03/2014). In 2008, apparently another electrification project financed by Rusal and managed by the government failed to become concrete (Interview with Activist 17/03/2014; Executive CBK 20/03/2014). Apparently due to this local corruption, but without doubt also because of the tense relations with the successors of President Lansana Conté, Rusal ceased paying taxes from 2007 to 2010 (Interview with Executive CBK 20/03/2014; Mayor of Mambia 18/03/2014; Sous-Préfet of Mambia 18/03/2014).

About 3 months before Lansana Conté's death, when the regime faced a democratic mass movement all over Guinea, this also prompted the inhabitants of Mambia to take the streets to demand for electrification, potable water, a health center, and a school. When people started to block roads and rails to stop the flow of bauxite, the military was sent in and killed and severely injured several activists (Camara 2008; Interview with Activist 17/03/2014; Haefliger 2012) . Many others were arrested, including the current mayor of Mambia, who was held in jail for 3 weeks (Sous-Préfet of Mambia 18/03/2014). Rusal finally gave way to the pressure from the streets and directly paid other operators who had been chosen by the commune itself to take over the electrification project (Interview with Administrator Mambia 20/03/2014). In contrast to the other company towns, Mambia thereby received a connection to the public power supply and the maintenance of the power poles remains within the responsibility of the public authorities (Interview with Mayor of Mambia 18/03/2014). In 2009, the money for the electrification was accounted for as "mining royalties" in the Mambian budget, although electrification is not part of the tasks of communes in Guinea (Interview with Administrator Kindia 24/03/2014). Most of the interviewees believe that Rusal treated these payments as taxes and thereby reduced its subsequent payments (Interview with Former Administrator Mambia 24/03/2014). Others—including the mayor of Kindia—claimed that this investment represented the accumulation of unpaid taxes of previous years (Interview with Mayor of Kindia 17/03/2014). A verification of these claims is not feasible due to missing local budgets and low transparency concerning the calculation of the taxes.

The contract between Rusal and the Guinean state of 2001 stipulates the payment of a royalty of one USD per ton, starting from the third year of commercial production (République de Guinée 2006: 22). According to the latest EITI report, CBK exported 3.2 million tons of bauxite for a value of 43 million USD in 2012 (Moore Stephens 2013: 57f) and paid taxes of altogether 32.3 billion GNF (ca. 4.6 million USD or 1.4 USD per ton) to the Guinean state in 2012. This is little compared to CBG, which paid 11.7 USD per ton in 2012 (cf. Moore Stephens 2013: 10, 34). An annex of the contract signed in July 2006 (Annex C) adds 0.1 USD per ton of exported bauxite as a contribution to local development starting from 2006 and a lump sum of 100,000 USD per year from 2001 to 2005 (République de Guinée 2006). This clear obligation concerning local development in Annex C is complicated by the fact that the assembly did not ratify it until 2011 and probably has not done so to date (Diallo et al. 2011: 52). Apart from the electrification project, the tax payments only started after the Alpha Condé came into power in December 2010.

During my interviews, none of the involved officials (neither civil servants nor employees of Rusal) was able to explain the basis upon which Rusal has actually calculated its taxes. I conducted interviews on this topic with the *directeur adjoint* of the mine of CBK, the *secrétaire général* of the trade unions in Débélé, the *directeur régional des mines de Kindia*, the mayor of Kindia, the sub-prefect of Mambia, with representatives of the community of Mambia, and several key activists of the civil society concerning mining in Kindia. Despite its online publication by the *Comité Technique de Revue des Titres et Conventions Miniers* (République de Guinée 2013b), none of these actors was in possession of the mining contract of CBK. Several civil society activists and the mayor of Kindia portrayed this transparency issue as the primary goal of the current movement (Interview with Activist 20/03/2014). The mayor of Kindia, Mamadou Dramé, pointed out that the local officials neither knew how much CBK exported nor did they possess the figures on the turnover of CBK:

"We would like to know [...] how much they have exported and on what basis they made their calculation. But today, we don't know that! [...] They say that in 2013 they gave what was due in 2012. This were 300,000 dollars, [...], but 300,000 on which basis? We don't know that. [...] Because of this financial manna there are nonetheless investments made by the CPD [*Conseil Préfectoral de Développement*; note by JK].

This is already good. But for things to become even better, we have to know how much has been paid and on which basis they made the calculation. Rusal must be able to tell the prefectural development council: this year, we have exported this amount of bauxite, a ton of bauxite costs that much, and the tax is the percentage of this turnover. It's zero point… what? We don't know! On this basis we have to calculate. It is 0.4%, but 0.4 of what? We don't know! The turnover has to be communicated!" (Interview with Mayor of Kindia 17/03/2014; translation by JK).

Dramé is also the president of the *Conseil Préfectoral de Développement* (CPD) and thereby a central figure in Kindia and Débélé. The CPD comprises an annual meeting of representatives of local and prefectural administrations and decides upon the distribution of the receipts from mining in the prefecture of Kindia. It does not take decisions concerning the actual disposal of the funds, but rather serves as an advising and supporting body (Interview with Activist 17/03/2014; Activist 20/03/2014). The CPDs had been introduced in the mining regions in September 2005 (Diallo 2005b: 8). According to Dramé, the investments are controlled by the local administrations themselves (Interview with Mayor of Kindia 17/03/2014).

The 300,000 USD mentioned by Dramé roughly correspond to the arrangement of Annex C of the contract between Rusal and the Guinean government in 2006. However, the fact that Rusal stopped paying shortly after and repaid several tranches at one time after Condé came into power complicates the picture. In addition, the new mining code should theoretically overrule the old contracts, although this also depends on the existence of the so-called stability clauses in these contracts. Last but not least, the new regulations also have to be implemented at the firm level by adapting the initial contract (Interview with Consultant 17/03/2014). In the case of CBK, this has not been done to date and it is unlikely that it will happen, given that the mine is only likely to exist for another 5–10 years before the reserves are depleted. The mining code of 2011 stipulates a "contribution to local development" of 0.5% of the turnover (République de Guinée 2013a: §130). The 3.2 million tons of exported bauxite led to a turnover of about 43 million in 2012 and would thereby lead to 215,000 USD of "contribution to local development". However, the *secrétaire général des collectivités* of the prefecture of Kindia registered a sum of 2,200,110,965 GNF (about 446,000 USD) in 2012, recorded as "contribution to local development" and "surface tax" (Interview with Administrator Mambia 19/03/2014). As the surface tax only amounts to about 17,000 USD (Interview with Activist 17/03/2014), this would mean that Rusal paid roughly twice the amount demanded in the mining code and thereby 1% of the turnover. By accident, this is the formula that applies for all minerals aside from iron and bauxite in the (new and the old version of) the mining code of 2011 (République de Guinée 2013a: 59, 2011b: 61) and this was also the formula mentioned by several informants (Interview with Activist 17/03/2014). Until recently, the situation was even more opaque, because Rusal did not publish its turnover. The activist Sita Mamoudou Condé described this situation as follows: "It's as if I tell you for instance: I give you 50% of what I have in my pocket. But you don't know how much there is inside" (Interview with Activist 17/03/2014; translation by JK).

3.1.2.2 Infrastructure and Corporate Social Responsibility

During the riots of 2009, a bus belonging to CBK was burned, and at the time of writing, it has not been replaced. With the same busses, CBK brings workers to the mines and pupils to their schools free of cost. Despite affirmations by the workers that they had not taken part in the riots, Rusal made the people of Mambia collectively responsible for the burning of the bus and did not replace it. As this measure had been an additional voluntary service of the company, the workers could not do anything against it. Some were compensated with 60,000 GNF (about 12 USD) per month, and since 2014, the pupils no longer need the bus, because a school has been inaugurated nearby.[95]

According to a booklet issued by Rusal, CBK has spent 1,742,766 USD on infrastructural projects since 2001. These projects were labeled as "contributions to local development" (Compagnie des Bauxites de Kindia 2013). The legal basis for these payments remains unclear. The administration in Mambia additionally listed a number of projects that did not enter the local budgets: since 2012, Rusal has constructed eight well drillings, two wards, two mosques, a secondary school, renovated the office (and at the same time residence) of the sub-prefect, funded women's groups, and donated for schools. In several cases of epidemics, Rusal also sent medicine. These voluntary measures outside the budget started with the coming into power of Alpha Condé, at the same time as the projects within the budget lines of the local government.[96] The bureaucratic processes behind these measures are quite simple: in one of the interviews, the mayor of Mambia told me that he simply wrote a letter to Rusal to ask the company to build a football ground. The company replied that it was going to construct a secondary school first, followed next by the football ground (Interview with Mayor of Mambia 18/03/2014). In Débélé, CBK finances a hospital ward for the whole population, sends about five busses per day in the surrounding areas to bring children to the secondary school of Débélé, organizes quarterly shopping tours to Kindia, allows the locals to hike on empty trucks,[97] organizes transport for sports events, and finances electricity for over 1000 inhabitants (Interview with Executive CBK 20/03/2014).

Housing has been a problem in Débélé since the start of production. Today, a substantial part of the workers commute between Kindia and the site, while others come from Mambia and other nearby villages. The rest live in either *Débélé cité* or the *Cité guinéenne*. *Débélé cité* also comprises the so-called *Cité pouchkine*, which is mainly populated by Russians and was described as a "cité dans la cité": an estate within the estate.[98] When I visited the *Cité guinéenne*, a well drilling was broken, which forced the inhabitants to seek for water in the pools and streams nearby. Employees of the facilities of CBK in Conakry receive primes for housing. However, in Débélé, workers without accommodation do not receive any compensation for their additional expenditure (Interview with Employee CBK 18/03/2014). Most of the infrastructure in Débélé can also be used by the surrounding population, even more as the area is not very densely populated. Both the medical ward and the secondary school—the *Collège de Débélé cité*—were built for the workers but are also used by the rest of the population, apparently free of cost (Interview with Administrators Mambia 18/03/2014). Until recently, the residents of Débélé were forced to go to Kindia to buy medicine. This changed in 2010 when a drugstore was authorized due to the pressure of the local trade unions. The drugstore belongs to a Moroccan who studied in Russia and is married to a Guinean CBK executive (Rusal 2010a: 8).

3.1.2.3 Expropriations

Most people who lived in the area of the current mine claim that they have never been compensated for their loss of land (Sous-Préfet of Mambia 18/03/2014). A worker, whose parents had been farmers in this area, stated that the affected farmers have claimed their rights since the mine has existed, albeit to no avail:

"Every time when they claim, they tell them to write letters and send them to the authorities. But every time they are writing, there are no benefits" (Interview with Employee CBK 18/03/2014; translation by JK).

Many of these farmers migrated to the center of Mambia, where they constructed at their own expense. In view of the near closure of the mine, these claims will become more important again. However, according to the officer in charge of agriculture at the sub-prefecture of Mambia, the land has no agricultural value anymore. The valleys—which would be sufficiently humid for cultivation—have been contaminated for too long by the mines above. Rather than conflicts with farmers, the company periodically encounters problems with cattle that graze in the mining areas (Sous-Préfet of Mambia 18/03/2014; Administrators Mambia 18/03/2014; Employee CBK 18/03/2014). According to Rusal, the new mine in Friguiagbé did not necessitate any relocations (Interview with Executive CBK 20/03/2014; Administrators Mambia 18/03/2014). To mitigate conflicts between the surrounding communities and the mine, the mining ministry also created a "Comité des mines", which is situated in Mambia and comprises the *directeur sous-préfectoral de la jeunesse* (DSPJ), the officer in charge of agriculture, representatives of women's organizations, and elders. They meet every 2 months and additionally in cases of urgency. Their mitigation not only concerns agriculture and environment but also working conditions and dismissals (Sous-Préfet of Mambia 18/03/2014).

3.1.2.4 Infrastructural Projects Under Control of Local Authorities

Local authorities in Guinean mining towns are permanently caught in between conflicting aspirations of the local population vis-a-vis the mining companies. They have to show their governmental authority whenever possible, prove that they are independent of the companies, and thereby, they are in a way forced to permanently put at risk the relations and thus the services and presents granted by the private actors. Several mayors of the mining towns—including the mayor of Mambia—emerged from resistance movements, were involved in violent conflicts, often incarcerated, and are now forced to cooperate with their former opponents. The ambiguity of this permanently displayed attitude of resistance became particularly clear in an interview with the mayor of Kindia, who is at the same time at the head of the body that coordinates the local use of the mining taxes and who reacted unexpectedly impulsively to the question of whether he had already been at the mining site. He negated and added that he did not even know what the mining director looked like:

"I, personally, have never been at the factory! [...] No, that's not bizarre, that's normal. Because if I go there—they are not part of my *commune*—what would that mean? What am I doing there? I don't know, to employ [sic]? I'm not the head of that *commune*! [...] No no no, there are no relations between them and me. Because if I go there today, for what would I do that? [...] No, for the moment there are not many things to work at together, but we will have to work together. That's what I've said: we have started a process! For the moment, they pay. They pass through the prefect who helps us and who makes pressure for that it comes. [...] There is no conflict whatsoever between us! Two or three years ago when the CPD was created, these people paid their arrears, on which basis I don't know" (Interview with Mayor of Kindia 17/03/2014; translation by JK).

In Mambia, three out of seven town councilors are former employees of CBK and the mayor had also been employed at the mine (Interview with Administrators Mambia 18/03/2014). The mayor had been a lorry driver at CBK before he got pneumoconiosis and finally retired in 2001. He is illiterate and does not speak French (Interview with Mayor of Mambia 18/03/2014).

In 2013, the prefecture of Kindia received local taxes of 322,746 USD. The amounts were promulgated in USD during a public ceremony. This was a great achievement, particularly as—according to the consultant Sandra Nichols—the mining code does not state anywhere that these funds have to be managed by the local authorities (Interview with Consultant 17/03/2014). The sub-prefectural administration of Mambia received 60% of this sum and is going to extend and build three schools, reforest 5 ha of land (that have not being used for mining), buy computers and motorbikes for the cashier and the *secrétaire general*, purchase furniture for the apartment of another officer, and extend electrification in a sector of the sub-prefecture. The latter again takes place despite the fact that—in legal terms—this is not part of the sphere of authority of local administrations. The last tranche had been used similarly, whereby the administration extended electrification and built schools and a health center. All of this was built in addition to the voluntary projects of Rusal/CBK, which have already been mentioned above (Interview with Administrators Mambia 18/03/2014). The remainder of the taxes goes to the prefecture of Kindia and other sub-prefectures around the mine (Interview with Activist 17/03/2014).

Unlike the procedures in Sangaredi and Kamsar, the local administrations actually control the funds for these projects. The cashier writes checks for the building enterprises that are signed by the mayor. This not only means that the democratically legitimized local council effectively controls its own tax incomes, but this also leads to more transparency for the population: "Here they tell you officially: this year, it's this amount of dollars. They tell you! [...] Over there, the people don't know [...] if its gifts or the tax" (Interview with Administrators Mambia 18/03/2014; translation by JK).

Besides the CPD, which is responsible for the distribution of the funds on the prefectural level and also serves as a supervisory authority, this local independence legally depends on the existence of the local development plans (PDL), which are compiled every 3 years (Interview with Administrator Sangaredi 30/01/2014) by a so-called *agent de développement local*. This agent is necessary because the local authorities do not yet have the required technical material (computers) and skills to do this according to the national standards. The computers ordered with the last part of Rusal's taxes had not yet arrived by the time of my visit. Altogether, the bureaucratic requirements for the PDLs only seem to be fulfilled in a rudimentary manner. The PDLs should be built on PDDs (*Plan de développement de district*), but these plans did not exist. The local budget is enacted based upon a *Plan annuel d'investissement* (PAI), which should be extracted from the PDL. However, no new PDL has been compiled since 2012. This means that the PDL will only be compiled after the investments have already been made (Sous-Préfet of Mambia 18/03/2014; Administrators Mambia 18/03/2014). Therefore, instead of being a democratic process starting at the district level and leading to long-term, forward-looking planning, the PDL rather remained an empty shell until the time of writing. Nonetheless, the administration is very proud of its new competencies.

3.1.2.5 Salaries and Working Conditions

Both working conditions and payment in Débélé are the lowest of the three bauxite mines in Guinea Employee CBK(Haefliger 2012; 18/03/2014). This has already been the case under OBK and SBK. According to Rusal, the average wage of CBK rose from 616,875 GNF in 2006 to 2,702,513 GNF during 2012 (Compagnie des Bauxites de

Kindia 2012).[99] If we take into consideration an average inflation of almost 20% during this time span (IMF 2014; World Bank 2015), this still leads to a rise in salaries of about 50%. However, most of the augmentation is due to a "bonus" of 40–50% of the basic wage, the so-called *prime d'encouragement*, while fringe benefits have been cut since the arrival of Rusal. The *prime d'encouragement* is only paid out if the production goal of the working unit (of about 20 workers) was reached and if the worker was neither ill nor on holidays during this month (Interview with Executive CBK 21/03/2014; Trade unionist 25/03/2014; Employee CBK 18/03/2014). This means that workers are not only punished for shortcomings of their colleagues, but they also lose one-third of their wage if they become ill or go on holiday. Without this disciplinary prime, net salaries have hardly risen since 2006. In addition, a large part of these primes seems to end up in the pockets of the senior management (Interview with Trade unionist 25/03/2014).

According to my own calculations based upon pay slips and salary scales, real net wages have most likely declined since the takeover of Rusal until 2006 and have only reached their initial levels again in recent years. In 1997, the base salary of an *ouvrier professionnel* (more than 50% of all employees) stood at 194,458 GNF, while boni were negligible. According to the salary scale of CBK of 2008, the base salary of *ouvriers professionnels* of about 500,000 plus a bonus called "condition de vie" amounted to about 630,000 GNF. Due to an inflation of about 15% from 1997 to 2008, this would have led to a decline in real terms of about 25%. With the *prime d'encouragement* salaries attained about 900,000 GNF in 2008, or quite accurately the same real wages as in 1997. If we consider that this prime is not always paid out and at least a part of it seems to get stuck in the higher ranks, as well as taking into account continuous cuts in fringe benefits, since the takeover of Rusal, real wages have actually fallen from 1997 to 2008. This is particularly surprising because the late 1990s were experienced by the workers as a time of crisis, mainly triggered by the aluminum wars in the former Soviet Union. In theory, the development of the living conditions in Débélé should have run quite parallel to the developments in Russia. Workers at Débélé, nonetheless, stated that they were content when production normalized again under Rusal (Interview with Employee CBK 18/03/2014).

Unfortunately, I did not find reliable data for the time between 1974 and 1997. Under the tutelage of the Soviets, wages were low but quite equal, medical services and schools were free of charge, and the workers were provided with food from the Soviet Union (sugar, rice, milk, potatoes, oil, and soap). After the death of Sékou Touré and the takeover of Lansana Conté in 1984, liberalization under the guidance of the IMF was also felt in Débélé. Since 1989, workers have to pay 20% of the costs of their medical services and pharmaceuticals. Only occupational diseases are fully paid by the company. Since 2012, Rusal only refunds generics instead of original pharmaceuticals (Interview with Trade unionist 25/03/2014; Compagnie des Bauxites de Kindia 2013: 3; Employee CBK 18/03/2014). In 2001, Rusal additionally stopped provisions in kind (Interview with Trade unionist 25/03/2014) and introduced 1-year contracts for 450 of previously 1650 core workers. Today, CBK only employs 800 core workers (Rusal 2015; cf. Executive CBK 21/03/2014).

The situation of the contract workers of CBK's main security company ESMUG is deplorable. Most of them have no contracts and earn 300–400,000 GNF (44–58 USD) as net salaries, without medical care. They have to endure working shifts of 24 h of work followed by 24-h rest, without holidays or vocational leaves. As in most of these cases, these workers are not unionized (Interview with Executive CBK 21/03/2014) and workers who complain are fired. One of the interviewed security guards spoke a very good French, was qualified for university entrance, and had a wife and children in Conakry. The family wanted him to come back, but the fear of not finding employment in Conakry made him stay as a guard at the entrance of the mine (Interview with Security guards 21/03/2014; Employee CBK 18/03/2014). According to the trade unions, Rusal does not respect the Convention collective mines et carrières signed by CBK and the Guinean trade unions in 2003 which regulates several boni in connection with occupational diseases, dust, dirt, education, and others that have not been implemented by CBK (Interview with Trade unionist 25/03/2014).

Promotions at CBK are rare. One foreman explained that after having replaced a Russian expert in 1995 who had been his superior for 10 years, he did not receive a higher salary than before. After more than 30 years of service, his seniority bonus only represents one-fifth of his net salary and he has never changed salary category (Interview with Trade unionist 25/03/2014). A trade union leader described this situation as a destruction of the status of the workers: "There is no career plan, there is only a production plan" (Interview with Trade unionist 25/03/2014; translation by JK).

At present, retirees receive small sums of about 50–120 USD every 3 months from the national pension funds. In 2001, Rusal added a one-off payment of 19 salaries.[100] Employees who started under CBK will receive better pensions (Interview with Mayor of Mambia 18/03/2014), although the aim of the trade unions remains to postpone retirement from 60 to 65 years. Since 1989, workers have to retire aged 60 and executives (cadres) aged 65 (Interview with Trade unionist 25/03/2014). Since 2013, CBK also pays retirees for reforestation measures to ameliorate their "social adaption" (Compagnie des Bauxites de Kindia 2013: 2; Administrator Kindia 17/03/2014). Officially, CBK prefers the plantation of fruit trees and species that favor the recovery of the whole ecosystem (Compagnie des Bauxites de Kindia 2013: 2). In practice, the company plants mainly one species that is particularly resistant. Neither the local authorities nor the persons in charge on the regional level seem to be implicated in the supervision of the reforestation (Interview with Administrators Mambia 18/03/2014; Administrator Kindia 17/03/2014).

3.1.2.6 Meager Prospects

Since the recent world economic crisis that hit Rusal particularly hard, the company has shut down several plants and suspended production in many others. Only in 2013, Rusal suspended production in the smelters of Volgograd, Volkhof, the so-called Ural smelters, parts of its facilities in Novokouznetsk and its Nigerian factory Alscon. Based upon its market power, Rusal hoped that with these measures, aluminum prices would start to rise again in the second half of 2014, while simultaneously complaining that certain producers—notably Chinese companies—foiled these efforts by continuously expanding their production (Deripaska 2014: 3; Rusal 2014e: 2f). The word crisis has also become part of everyday language in Débélé, even if production has, nonetheless, been augmented ever since (Guinée Emergence 2014) and the company has also invested in 40 new rail cars, several new locomotives, and new mining machines (Interview with Trade unionist 25/03/2014; L'Observateur 2014: 9; Soloviev 2011: 2). This is all the more surprising as the reserves in Débélé are almost depleted. Algassimou Diallo, the vice director of the mine, expects another 6 years of production, after which all lucrative reserves will have been exploited. In theory, there would be reserves for a few more years at a distance of about 20 km from the current mines, although this investment is not considered profitable at present (Interview with Executive CBK 20/03/2014; Executive CBK 21/03/2014; Administrators Mambia 18/03/2014). Despite the deplorable situation in Fria, neither Rusal nor the political authorities have plans for the future of Débélé after the mine is depleted. Algassimou Diallo asserted that the workers will be sent to Dian-Dian afterwards, where exploitation could theoretically last for another 100 years (Interview with Executive CBK 20/03/2014; Chambre des Mines de Guinée 2011: 33). While this relocation of the workers is already problematic in itself, it would also lead to a severe degradation of the infrastructure and the impoverishment of the whole sub-prefecture. Hundreds of subcontractors, merchants, taxis drivers, craftsmen, and many other service providers would either be forced to follow the workers to Dian-Dian and hope to find a new niche for them there or face poverty in Débélé. According to Ibrahima Diallo, the company already started to dismantle workshops in Débélé in August 2014 to re-erect them in Dian-Dian (Diallo 2014r).

3.2 The Compagnie Des Bauxites de Guinée

The *Compagnie des Bauxites de Guinée* (CBG) is the largest African bauxite producer and is currently owned by the Guinean state (49%), Alcoa (22.95%), Rio Tinto Alcan (22.95%), and the Swiss company Dadco S.A. (5.1%). It supplies almost half of the bauxite imported to the US and Canada (Committee on foreign affairs 2007: 41) and the vast majority of the bauxite imported to Germany (Jolly 13/01/2015). Dadco has its head office on the tax heaven of Guernsey, one of the Channel Islands, and its sales office in Lausanne, Switzerland. It has no production sites in either of those countries. Apart from its share in CBG, its only industrial asset seems to be the Aluminium Oxid Stade GmbH (AOS), a refinery in Germany that receives its bauxite from CBG (Jolly 13/01/2015). The further destinations of Dadco's alumina are kept secret by the company (Jolly 13/01/2015), but it is very likely that it is processed in Germany. Dadco acquired a 50% share of the refinery in 2001 and fully bought it in June 2004. On its homepage, Dadco states that it is "part of a larger group of companies; the founding company was established in 1915" (Dadco 2013). However, no further indications are given about this larger group of companies or the founding company.

Fig. 3.6 Advertisement of Rio Tinto in Conakry, February 2012 (JK)

Despite this rather dubious company structure, this firm was very helpful in establishing contact to its bauxite supplier in Guinea. When I contacted this last German alumina refinery in Stade, I was forwarded to a representative of Dadco, who very kindly arranged a stay in Sangaredi, surprisingly without any objections, bureaucratic requirements, or questions about the exact purpose of my stay. The reception by the director of the mine in Sangaredi was also very generous, but at the same time as curt as one would imagine the encounter of a mining manager with a social scientist.

Rio Tinto started exporting Guinean ores only in 2007 when it acquired Alcan (which had acquired Pechiney in 2003) and thereby became part of the new premier league of aluminum, alongside with Alcoa, Rusal, and Chinalco (Fig. 3.6). Together with its other assets in copper, iron, diamonds, coal, uranium, and gold mining, it also ranks among the three largest mining companies in the world. In 2007, BHP Billiton—then still an important player in Guinea—tried to take over Rio Tinto. The result would have been by far the largest mining company worldwide. This led to numerous reactions by both governments and other mining companies, which finally prevented this merger. Among them was Chinalco, which bought a 9% share of Rio Tinto and thereby became its largest shareholder. According to Custers, this move was entirely unexpected (Custers and Matthysen 2009: 47). Rio Tinto has already been present in African countries for at least half a century (Lanning and Mueller 1979: 334). The Guinean bauxite of Rio Tinto most likely goes to Gardanne (France) and Saguenay (Quebec). From there, the alumina goes to smelters in Canada, France, and probably also Iceland.[101] Rio Tinto's smelter in Lynemouth had initially been part of a production network with Fria and Aughinish, which are now owned by Rusal (Campbell 2009b: 79; Deguenon and Traoré 2011a: 4; cf. Rio Tinto Alcan 2015).

Alcoa is one of the oldest aluminum producers worldwide and the last influential "survivor" of the Six Sisters. It wielded most power from the 1940s until the 1970s, although it remains among the largest four producers, operating in over 30 countries on all continents. In Africa, Alcoa controls factories in Guinea, Morocco, and South Africa. Its share of Guinean bauxite goes to a refinery in Point Comfort, Texas, as well as San Ciprian, Spain, where both alumina and aluminum are produced for customers in Europe, Asia, and North America (Harden 17/11/2014; cf.Campbell 1986: 57; Diallo 2014g). This includes the aircraft industry (Interview with Journalist 15/01/2014), car companies, packaging, building and construction, commercial transportation, and consumer electronics (cf. Alcoa 2015). Both Alcoa and Rio Tinto are mainly producing based upon long-term supply contracts with major processing companies, such as Ford and General Motors. Only a few big aluminum companies have the means to do that (Brinkmann 2004: 17).

The control relations in Sangaredi (Bauxite mining) and Kamsar (Bauxite concentrator and ship loading) are similar to those in Fria and Edéa, although CBG has not yet received as much academic attention as Pechiney. A Canadian employee in Kamsar called CBG a "mini society"(Interview with Executive CBG 20/02/2014). According to this employee, the company is primarily a bauxite mine, but it is also

"a railroad, a refinery, a harbor, two cities to accommodate the workers, the city of Kamsar and the city of Sangaredi, a sewage treatment plant, a drinking water plant, two power plants to fuel [the cities, but also] accommodation centers, restaurants, clinics, [...] the hospital of Kamsar, it's a vehicle fleet", etc. (Interview with Executive CBG 20/02/2014; translation by JK; see also Soumah 2008: 141).

The vehicle fleet also includes cranes and a tow car and the company even maintains the only morgue in Kamsar. This need to "provide virtually everything"(Interview with Executive CBG 20/02/2014) is sometimes difficult to explain to the parent company, which normally relies much more on public services like electricity and the maintenance of infrastructure:

"We often have to explain our partners that in the Guinean context, because you have a government which is... very poor and which offers very little service... in this moment it's our task to make up for this lack" (Interview with Executive CBG 20/02/2014; translation by JK).

For instance, cranes are not needed very often, but if a train becomes derailed, they are indispensable. The supply of water and electricity creates similar problems as in Fria. Even the prefectural capital Boké, situated about an hour's drive to the East of Kamsar, receives electricity and water from its facilities (Interview with Executive CBG 20/02/2014). When the population of the district Bendougou protested against the waste deposal practices of CBG in 2014 by blocking the roads for 5–6 months, the company stopped the waste deposal for the whole city. As the local administration was overburdened by this task, the NGO CECI—also co-financed by the companies—took over by sponsoring new waste dumps (Interview with Sous-préfet of Kamsar 12/02/2014; Executive ANAIM 17/02/2014). In the 1990s, due to pressure by the Word Bank, CBG started to "privatize" (Interview with Executive CBG 18/02/2014) considerable parts of this never-ending list of facilities. This concerned among others the only bakery in town, the cinema, several hotels and clubs, agricultural firms, and schools. The latter have also been outsourced due to the shrinking of expatriate personnel (Interview with Executive CBG 18/02/2014; Executive CBG 20/02/2014).

With a turnover of 500 million USD (République de Guinée 2013c) and annual payments to the government of about 160 million USD (Barry 2013b: 7)—compared to about 4 million of Rusal (Interview with Legal expert 13/03/2014)—CBG is often described as the most influential organization in Guinea, whose leaders are part and parcel of national politics (Interview with Administrator Sangaredi 3/02/2014; Former executive Rusal 3/04/2014). The production price of bauxite in Kamsar is quite high. It has been relatively stable between 20 and 25 USD per ton since 1995, compared to 10–15 USD at CBK

(International Monetary Fund 2008: 9). As the shareholder structure shows, private investors hold the majority of CBG and have thereby the last word in the board of directors, comprising six government appointees and six representatives of the involved firms. In case of a standoff, the managing director—himself not belonging to the board—is allowed to vote (Soumah 2008: 129; cf. Compagnie des Bauxites de Guinée 1983). In 2010, CBG got its first Guinean managing director, Kémoko Touré, whereas all the previous directors had been expatriates from the US, Canada, Australia, France, and Great Britain. During my stay, Alpha Condé was reported to have proposed Kémoko Touré as the new prime minister (Diallo 2014u). The headquarters of CBG in Kamsar—coincidentally or otherwise—are called the "White House" (*Maison Blanche*) by the population.

However, this central institution, being responsible for most of the exports of Guinea and vital to the government budget, only counts 2300 employees. By contrast, Guinean artisanal gold mining provides jobs for 50,000–100,000 people (Chambre des Mines de Guinée 2011; République de Guinée 2013; cf. Soumah 2008: 51, 138). This also has considerable consequences for the mining towns. In 2011, the mine in Sangaredi employed 504 workers of CBG, 629 "permanent contract workers", 100 "seasonal contract workers" and maintained 489 houses for its workers, among a total population of about 75,000 (République de Guinée 2014: 7). In Kamsar, a town of 113,000 inhabitants, the factory directly employs less than 2000 people.[102] Even if we suggest that every worker provides for 20 other persons, this leads to a much larger part of the population being both excluded from and dependent on the productive centers, in terms of social services, infrastructure, and even security.

This tendency is reinforced by the marginal spin-off effects of aluminum production. CBG sends the mined bauxite from Sangaredi to Kamsar, where it is only prepared for export to a wide range of ports in Europe and the US. CBG has two procurement agencies—in Brussels and Pittsburgh—that also receive the demands of African facilities. They very rarely order products from Guinea or other African countries (Interview with Executive CBG 14/02/2012). One exception could be the provision of the workers with agricultural goods, although even this possibility to create more value added and more jobs in Guinea has not been put into practice to date, with CBG blaming a lack of stable providers of these commodities (Interview with Consultant 10/02/2012).

This export enclave character enforces the development of the mines into separate sovereignty regimes.

3.2.1 History

The history of CBG also stretches back to colonial times.[103] From 1920 to 1958, *Bauxites du Midi*—the predecessor of CBG—explored the bauxite reserves in the Boké region (Compagnie des Bauxites de Guinée 2012; Pezet et al. 2009: 19). The current military camp in Boké had previously been the camp of *Bauxites du Midi*. From 1948 to 1965, the company already shipped bauxite from small mines from the Loose Islands (Tamara and Kassa) near the capital to refineries in Canada (Campbell 1983: 67; Ribadeau-Dumas 1952: 18; N'Diaye 1984: 25).[104] In 1957, the Canadian company Alcan bought the firm from French owners,[105] formed a consortium with Alcoa, Kaiser, Reynolds (all of them US), Alusuisse (Switzerland), and VAW (Germany) and already began to build a bridge over the Rio Pongo, mainly financed by Alcan itself (Soumah 2008: 123). However, in November 1961, the government of Guinea nationalized the mines in Boké and Kassa, because Alcan did not respect a former agreement to build processing plants in Guinea.[106] This was the first nationalization of an aluminum facility by a non-industrialized state and served as a role model for similar developments within the framework of the International Bauxite Association 10 years later (Holloway 1988: 42; Forster 1976: 135). How this nationalization actually took place and how Alcan was compensated remains unclear (Soumah 2008: 147). In any case, Sékou Touré acted quite pragmatically compared with Guyana,[107] which nationalized its mines 10 years later on the eve of the foundation of the International Bauxite Association. Touré immediately started negotiations with the US junior company Harvey, mediated by the US ambassador in Conakry, William Attwood.[108] At that time, the US government under J.F. Kennedy struggled to reduce the influence of its former monopolist Alcoa and, therefore, embraced this opportunity to combine anti-communist containment policies with the safeguarding of valuable reserves for its booming aluminum industry. The Kennedys apparently maintained good relationships with both Lawrence Harvey—the head of the company—and Sékou Touré, relationships which were established before Kennedy came into power.[109] According to Holloway, Harvey was a typical "fringe" company, i.e., small firms that did not succeed in challenging the aluminum oligopoly until the 1970s. Its joint action with a post-colonial socialist regime was a typical strategy, which was also pursued by the petroleum industry (Holloway 1988: 45). Prior to entering Guinea, Harvey acquired a smelter from the American Defense Plant Corporation, which had been privatized after WWII, but mainly due to the resistance of the Six Sisters, the firm was unable to establish a sustainable production chain. In 1963, the company's fortunes finally seemed to change when Sékou Touré chose Harvey as a joint venture partner (49% Guinea and 51% Harvey, while 65% of the profits would belong to Guinea) for the nationalized mine (Holloway 1988: 33–36; Soumah 2008: 125). However, in this case, the capital of Harvey was apparently too small to stem the project alone. According to internal accounts of CBG, the decision makers of Harvey and the Guinean government soon realized that the planned output of CBG of 1 million tons would not suffice to pay back the assumed debts (Johnson-Sahr 1991: 3). Other accounts rather point to the counter strategies of Alcan and other major companies that had been excluded from this promising mining project (Soumah 2008: 123; République de Guinée 2013c: 9). This combined pressure of a large part of the oligopoly led to intense negotiations from 1963 to 1966. Ultimately, the World Bank (Guinea had been admitted to the IBRD in 1962) and USAID only agreed to co-finance the project under the condition that Alcan, Alcoa, and Pechiney were brought back into the game.[110]

Ibrahima Soumah, who was mining minister and thereby also chair of the board of directors of CBG from 2000 to 2002 (Bah 2013), also mentions that the World Bank president from 1963 to 1968, George D. Woods, had been working as a financial advisor for another US aluminum major—Kaiser Aluminum—before becoming World Bank president (Soumah 2008: 146). During his time in office, this company was financed by the World Bank to build the enormous Akosombo dam (finished in 1965), which provided the energy for another smelter. Both projects triggered violent clashesat the time of their completion. Kwame Nkrumah, the first president of Ghana, was overthrown only a few days after the ceremonial opening of the Akosombo dam on 22 January 1966 (Lanning and Mueller 1979: 429–435; Wiederstein 1994: 37) and the Portuguese invasion of Guinea on 22 November 1970 took place only 2 days after a CBG board meeting at the *Case de Belle Vue*, where an agreement on capacity expansion had been signed (Soumah 2008: 150). This villa was then burned down by the Portuguese forces during their unsuccessful search for Sékou Touré (Marx 1977: 4). The invasion (Operation Green Sea) had been supported by a considerable number of Guineans and also led to "massive" arrests of Guinean staff at CBG (Soumah 2008: 150).

In 1968, Harvey was bought out by yet another US company, Martin Marietta (Holloway 1988: 33–36). Beside the Guinean government, which still held 49%, CBG was then held by the consortium Halco mining (51%) which in turn belonged to Alcoa (US, 27%), Alcan (Canada, 13.77%), Martin Marietta (US, 27%), Pechiney (France, 10%), VAW (Germany, 10%), and Montecatini (Italy, 6%; Soumah 2008: 128). Until today, CBG maintains that the plateau of Sangaredi is the richest and the largest in the world and that the aluminum content of the Sangaredi reserves (60–70%) was never to be found again (Interview with Employee CBG 14/02/2012).[111]

3.2.1.1 OFAB and the Start of Construction

With the start of constructions in the late 1960s, the "mangrove swamps" (Compagnie des Bauxites de Guinée 1983: 13) of Kamsar and the "plateau covered with grass" (ibid.) that was to become Sangaredi began their metamorphosis into crowded mining towns (Interview with Retirees Bauxites du Midi 12/02/2014). Two aged councilors in Kamsar described the working conditions in these building lots as being particularly harsh. Wages were low and the builders were forced to work on weekends (Interview with Retirees Bauxites du Midi 12/02/2014; Johnson-Sahr 1991: 7–9). Following the example of MARG in Fria, the Office d'Aménagement de Boké (OFAB) was formed in December 1965 to build and maintain the infrastructure of the project, namely, the harbor, the railway, and the cities of Kamsar and Sangaredi. Similar to MARG, OFAB was financed by external donors—this time by the World Bank and USAID—and received technical assistance from the Belgian firm Tractebel (Soumah 2008: 113–115). OFAB served mainly as a link between the major donor—the World Bank—and the sub-contracting companies in charge of building the infrastructure. It was controlled by the ministry of "economic development" (Compagnie des Bauxites de Guinée 1983: 10) and had the explicit mission to prevent the development of an enclave. Both this mission and its powerful position as a broker made this unit very dominant and thereby led to permanent conflicts between the government and the private stakeholders, for instance, about the "nepotistic" employment practices of OFAB (Interview with Executive ANAIM 21/02/2014; Retirees Bauxites du Midi 12/02/2014; Ibrahima Sylla in Soumah 2008: 120). A suppression of the workforce comparable with Fria did not take place. Touré seemed to support strikes up to a certain point, albeit not without "forgetting the profitability of the factory" (Interview with Trade unionist 28/03/2014; Journalist 28/03/2014). In 1983, CBG called this cooperation "a landmark event in world affairs. Consider: at a time when many thought it impossible to reconcile socialism with private enterprise, a nonaligned socialist republic—Guinea—entered into direct ownership of a joint business venture with a private firm" (Compagnie des Bauxites de Guinée 1983: 10). Confronted with this question, a senior executive of CBG (ANAIM) stated the following:

"[CBG had] nothing to do with the socialists! [*laughs*] Do you understand? The basis, the great factory, the construction that we have made here, during the first [government], this was based on the approval and the benediction of the American government, the Canadian government and the French government. The socialists have made something apart. [...] This is to say that this had nothing to do with the regime. Absolutely nothing" (Interview with Executive ANAIM 21/02/2014; translation by JK).

The senior executive continued to state that Kamsar worked out well because the government did not interfere. To Guineans at this time, "Kamsar, that was America!" (Interview with Executive ANAIM 21/02/2014). This external dependence was deepened by increasing foreign debts (cf. Dux 2000; Legal expert 13/03/2014). Rather than guaranteeing local processing and spin-offs in Guinea as provided by the contract, the high state participation in CBG thereby "appeared with time as perfectly compatible with the logic of accumulation of transnationals involved" (Campbell 1986: 31; cf. Shafer 1986: 925). Soumah showed that the public–private management of CBG has a private nucleus until today. The chair of the board of directors is the Guinean minister in charge of mining, but this position is "more formal than real" (Soumah 2008: 129), mainly because his corporate counterpart—the managing director of CBG—is able to take decisions in the case of standoffs.

In 1995, the OFAB was replaced by the *Agence nationale d'aménagement des infrastructures minières* (ANAIM), whose mandate was enlarged while its influence in Kamsar was reduced. The organization was now responsible for the infrastructure of all mining enterprises in Guinea (Interview with Trade unionist A 25/02/2014). While OFAB had a direct say at CBG, ANAIM now rather acts as a "watchdog of the government" (Interview with Executive ANAIM 17/02/2014) and aims at a coherent infrastructural planning for all mining enterprises in Guinea; for instance, by planning future railway lines that could serve several mines and factories at a time while additionally providing public transport. In Kamsar and Sangaredi, ANAIM owns the railway lines, the hospital, the port and half of the buildings of Kamsar Cité, but it does not control them. Only the hospital is also managed by ANAIM, despite, nonetheless, being funded by CBG. The rest of the infrastructure is rented out to CBG, which is in turn responsible for maintenance. ANAIM, nonetheless, serves as a broker between the state, CBG and other mining companies in case of demands for the existing infrastructure (Interview with Executive ANAIM 21/02/2014). The salaries of ANAIM's about 300 employees stem from its infrastructure rents from CBG, except the personnel of the hospital, who are *more* directly paid by CBG (Interview with Executive ANAIM 17/02/2014). Most of the employees of ANAIM still work in Kamsar (Interview with Hospital director 19/02/2014; Executive ANAIM 21/02/2014). A young officer at ANAIM gave me a brief account of his daily work at ANAIM, which seemed to be surprisingly varied. During his working time, he also developed projects that were only loosely connected to mining (like reforestation in areas that had never been touched by mining projects or campaigns against illegal fishing) and he even administered his own environmental NGO (Interview with Executive ANAIM 17/02/2014). The control structure of the ANAIM hospital is rather complicated. According to CBG, the hospital—and thereby ANAIM, which in turn reports to the mining ministry (Interview with Hospital director 19/02/2014)—receives annually about 3.5 million USD for its functioning (Interview with Executive CBG 20/02/2014). During one of my stays in Kamsar in February 2014, the whole ANAIM personnel was on strike due to outstanding wages (Interview with Executive ANAIM 17/02/2014; Police Commissioner 17/02/2014).

3.2.1.2 The Start of Production

Production in Sangaredi and Kamsar finally started in 1973. While most of the bauxite was directly shipped to the US in the 1970s, new facilities in Ireland and Spain led to an increasingly important participation of European factories in the production chain. This was contradicted by

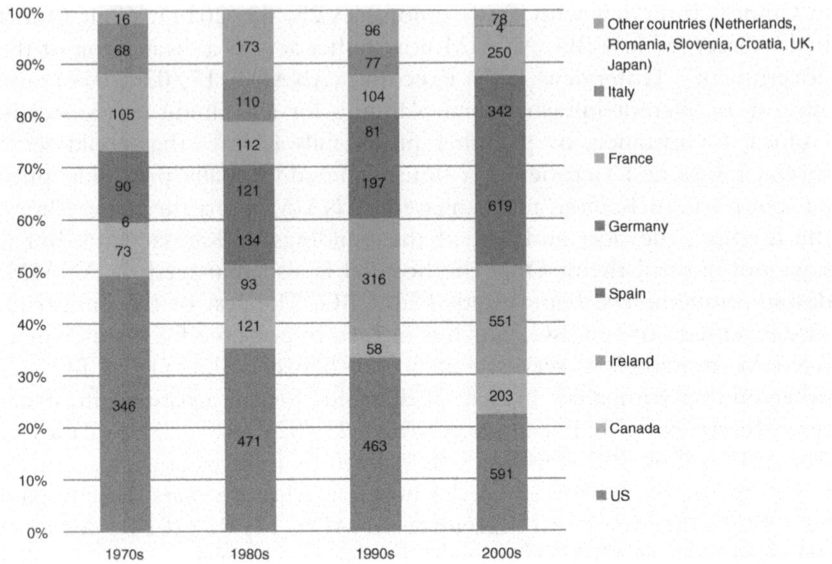

Fig. 3.7 Number of bauxite boats of CBG by countries of first destination 1973–2011

a continuous fall of shares of European firms. The laborious evaluation of more than 6000 chemical analysis sheets of bauxite ship loads that I found in the archives of the mine of Sangaredi show that the first destination of Guinean bauxite boats has changed drastically during that last 40 years of production. The share of American (US and Canadian) ports fell from almost two-thirds to less than one-third (Fig. 3.7).[112] At the same time, the ownership of American firms at CBG rose from 38.25 to 45.9%.[113] From this perspective, the recent raw material crisis in Europe is not a crisis of underproduction but a crisis of control over existing (mines and) refineries in Europe.

Only 1 year after the start of production, Alcan announced that its part of the bauxite was not to be locally processed as previously announced, but rather in Aughinish, Ireland, which in turn was to provide alumina for a smelter in Lynemouth, UK. This change of mind was obviously part of a general power struggle between newly independent governments and the old capitalist centers, which culminated in the first oil crisis and also led to the formation of the IBA in 1974 (see Sect. 2.5).

Campbell remarks that it "would be quite impossible to explain the logic of the Aughinish project in terms of the comparative costs of the factors of production" (Campbell 2009b: 79), despite extensive concessions made by the Industrial Development Authority in Ireland to attract the investment (Campbell 1983, 1986: 10, 49; Campbell and Clapp 1995: 428). Only 3 years after the start of production in Aughinish, in 1983, Alcan threatened to abandon the project again if production costs could not be brought down in both Aughinish and Kamsar (Campbell 1986: 10, 49).

Sékou Touré's vision of CBG was an integrated industrial complex that would reach out to other sectors and thus create growth in Guinea:

"The struggle for economic and social progress will depend on our resolution to increase the intellectual and technical capabilities of every Guinean. [...] Planners are looking to a synthesis between agriculture and industry. Industrial growth will help boost agricultural output by providing farm machinery, synthetic fertilizers and other useful products. Conversely, agriculture will provide the raw materials for more industrial growth. Crops like palm, sugar and cotton are "industrial" crops: they're the base for processing, refining and textile industries." (Compagnie des Bauxites de Guinée 1983: 38).

In the outcome, this "synthesis" was rather reduced to CBG financially contributing to the erection of agroindustrial complexes. The company itself was not connected to these complexes. In addition, funds did not flow as desired by the government in the beginning. The contract of CBG provided that 65% of its profit was going to the state despite the latter only holding 49% of the stakes (Compagnie des Bauxites de Guinée 1983: 14; Soumah 2008: 125–128). However, despite sharply rising aluminum prices (Campbell 1986: 9), CBG did not declare any profits for several years after the start of production. This led to considerable "frustration" as the infrastructure of the project now figured in the debts of the Guinean state (Soumah 2008: 171). CBG explained this with the fact that it had just started production. However, due to the vertical integration of the participating companies, it would have been easy to circumvent the payment of dividends to the Guinean state (for transfer mispricing see Bauer, Maissen 1989: 92–101, 132f; cf. Soumah 2008: 176). Similar practices have been reported in connection with Anglogold Ashanti operating in Siguiri (SAG). Despite 15% state ownership, the government has only recently received its first dividend, apparently because the company had always declared a deficit. Today, the company pays almost one-third of all governmental income from mining and is thereby the second largest source of government revenue after CBG (Interview with Legal expert 13/03/2014; Moore Stephens 2013: 10).

This led to the so-called special tax, which passed parliament on 13 January 1975 and was applied to exported bauxite (0.5/ 0.75% of aluminum ingot),[114] alumina (1% of aluminum ingot) and also iron (1 USD/t). Due to its comparatively little output, Fria only contributed about 5 million USD to the budget in the following years. The taxes paid by CBG—after "long-lasting and difficult" negotiations—were considerable. From 1977 and until the 1990s, the government received between 150 and 200 million USD per year from CBG. In real terms, this is almost twice as much as all governmental income from mining today.[115] The bauxite prices themselves were pre-arranged, based upon a formula that included inflation in the US, petrol prices, aluminum prices and labor costs.[116] From 1975 to 1984, minerals—mostly bauxite from Sangaredi—were responsible for 90–95% of exports. After independence, agricultural products still made up about 70% of exports (Dux 2000). One interviewee called this situation a revolution on the payroll of capitalism:

"CBG has always been the company that has truly paid a lot compared to the others. We have lived on the back of CBG. [...] It's the only case where I would say that the state has succeeded to tear out something that yields. Everywhere else this is not evident. For CBG, this is clearly visible to the naked eye." (Interview with Legal expert 13/03/2014; translation by JK).

The Africanization of CBG did not proceed as quickly as was foreseen. In 1973, 170 expatriates were employed in Kamsar and had to train the Guinean senior staff (Compagnie des Bauxites de Guinée 1983: 5). However, according to Soumah, this training did not take place or was obstructed, because the expatriate trainers themselves had no guarantees to retain their jobs after successfully passing on their know-how (Alfa Issa Thiam in Soumah 2008: 142; for similar problems in Fria see Larrue 1997: 243). In addition, most of the Guinean specialists who were initially sent abroad for qualification did not come back. To reduce this brain drain, the government founded the *Institut supérieur des mines et de géologie de Boké* in 1972, which trains most of the future employees of CBG until present (Shundeyev 1974: 38). Due to these obstacles, the initial aim of CBG to replace all the expats by Africans within 5–10 years has not been reached until today. In 1997, Dux counted 100 “foreign experts” (Dux 2000). The first Guinean director of CBG only assumed his post in 2010. Soumah also relates this delay to the strong nepotism in the recruitment of Guinean personnel, which reduced the possibility to employ skilled workers (Soumah 2008: 212).

The same author also criticized the privileged situation of expats, who received much higher wages and were granted many fringe benefits that Guineans did not have (Soumah 2008: 212; Employee CBG 7/02/2014). These privileges also concerned housing and the use of certain facilities, prompting Alfa Issa Thiam—an employee at CBG/OFAB—to speak of “total apartheid” in Kamsar until a change of general managers in 1977. When a son of Sékou Touré, Mohamed Touré, was apparently denied access to the company’s swimming pool due to his color in the late 1970s, these practices finally seemed to attract the attention of the revolutionary regime (Alfa Issa Thiam in Soumah 2008: 143).

After a boom in the 1970s, the aluminum sector was strongly affected by the slowdown of worldwide growth in the 1980s (Campbell 1986: 9). The reaction of the majors was comparable to the recent crisis: they complained about the continual expansion of productive capacities in third-world countries but failed to considerably reduce capacities themselves to stabilize the price. From the perspective of a single company, this behavior is logical, because reducing capacity during a crisis by shutting down whole plants also means giving up market shares after the crisis, particularly due to the long timespan necessary to build up a mine, a refinery or a smelter (Campbell 1986: 9). In the 1990s, despite a decline in bauxite prices, the nominal export revenues of the Guinean bauxite sector were fairly stable, because the exports of CBG increased slowly but steadily. However, as already mentioned, tax revenues had, nonetheless, been reduced by half in real terms since the 1970s, mainly due to changes in the tax regime in CBG, but also because of financial problems of Fria and CBK (International Monetary Fund 2008: 9). In 2000, shortly before the new boost in raw material prices, CBG negotiated another reduction of taxes (International Monetary Fund 2008: 9).

3.2.2 Kamsar

Kamsar is the largest bauxite town of Guinea and is often depicted as the economic capital and even as a shadow government of Guinea. The working conditions are relatively good and the overall situation of the city is much less problematic than in Fria and Débélé. When the former general director of CBG, Kémoko Touré, was replaced by another Guinean, Namory Condé, he nonetheless heavily criticized his predecessors:

"In fact, the CBG was not in a very good condition. There are multiple causes. I would mention the following: a total absence of vision [...]. A neglect of living and working conditions and of salaries. An erosion of accumulated know-how for decades due to the multiplication of local and external sub-contracting accompanied by suspicions of active or passive corruption by certain associates. [...] An atmosphere of widespread theft and of conflicts of interests." (Barry 2013b: 7; translation by JK).

Despite this quite negative presentation of the company by its own head, the working and living conditions of the 2250 permanent employees of CBG (all facilities; Chambre des Mines de Guinée 2012: 28) can still be considered among the best in Guinea. According to Soumah, the wages in the mines are about four times higher than in the public administration and are accompanied by "modern" housing with electricity and tap water, free medical care for the family, sport and cultural centers, basic food provisions and school libraries (Soumah 2008: 213). The food provisions comprise US rice, Austrian sugar, sweetened can milk from Germany, canned tomatoes from Italy, English cooking oil and Belgian soap. All of these goods are provided through the so-called *Provision Alimentaire Mensuelle* or PAM (Interview with Employee CBG 7/02/2014; Security guard 7/02/2014). During the start of operations, this list even included bread, tinned fish and canned meat (Interview with Retirees Bauxites du Midi 12/02/2014). Most of the workers' families consume all these products themselves. Particularly the rice rations—ranging from 45 to 135 kg per month—are consumed very quickly and are often given out in advance and then deducted from the wages. The extent of the provisions depends on the number of children per family, apparently with a ceiling at the fourteenth child. The provisions are limited to one parent per family. Concerned interviewees argued that this results in a lower real wage of the second (or third) partner if he or she is also employed by the company. This also leads to problems if the husband marries another wife. As men can marry up to four wives, they normally do not get divorced and sometimes neglect their older wives and children. Workers without families (or in need of cash) sell their goods on a well-established black market (Interview with

Employee CBG 7/02/2014; Security guard 7/02/2014). Moreover, a large number of state employees (policemen, teachers, administrators, etc.) receive food provisions from CBG (Interview with Administrator Sangaredi 3/02/2014).

The basic salary of the lowest category of permanent workers amounted to 1,851,142 GNF or about 270 USD in 2014, compared to 162,000 GNF in 1993, which would today amount to about 972,000 GNF or 140 USD (Interview with Trade unionist 27/01/2014). This is about 1.7 times higher than the lowest basic salary in Fria (figures for 2010 in both cases). Wage demands have sometimes evolved into strikes (Diallo 2005d: 5), although the trade unionists seem to have better relations with both the government and the companies compared to Fria. A small part of the upper ranks of CBG employees enjoys living conditions that can be compared to their colleagues in Europe or the United States. Some of their children are schooled abroad and they acquire real estate, businesses or firms in Guinea to sustain their living standards after retirement (e.g., Executive ANAIM 13/04/2014; Unemployed 20/02/2014).

However, the relatively good living standard inside the protected realm of the company town can quickly transcend into poverty if the workers' (extended) families expand too quickly. In addition, the borders of the CBG "polity" are quite tightly drawn both spatially and temporally: non-permanent workers of CBG mostly live outside the *cité* without tap water or electricity and often do not even receive the national minimum wage of 440,000 GNF (about 60 USD; Vision Guinée 2013; Administrator Kamsar 10/02/2014). At the time of their retirement, the core workers also lose all their privileges and are only left with the low allowances of the state pension fund. The latter is also used for evacuations, accidents and family allowances and has been insolvent for several years since 2006. Already in September 2005, workers and their partners pushed for higher pensions by blocking the mines (Diallo 2005a: 8). The insolvency also related to the shrinking number of core workers, which significantly reduced the already fragile national pension fund.[117]

According to Mamadouba Soumah, the general secretary of the *Fédération syndicale professionnelle des mines, carrières, industries chimiques et assimilées*, out of the 1320 workers to have retired between 1996 and 2005, 200 have died under "conditions of extreme poverty" (Diallo 2005d: 5). This happens if workers have not accumulated sufficient savings before their retirement and do not have a successor at the factory. My host family in Kamsar faced comparable problems. When I arrived in Kamsar, I was welcomed by the family of a train operator of CBG who was related to acquaintances in Conakry and Sangaredi and felt obliged to accommodate me despite my affirmations that I had found a suitable hostel. After a long conversation during which the potential host asked me if I was used to “African” conditions, I did not know if the family was rather discomforted by the idea that I would spend the night there or if they would rather be vexed if I refused. Eventually, I agreed and found myself sleeping in the nursery, in a short bed that was so worn that one crossbar in the middle of the bed directly touched my spine. My host was in the last year of his service and had two wives but only seven children. Nonetheless, these children had already children themselves, which led to about 26 people regularly eating with the family or receiving parts of its subsidized rice (Interview with Worker CBG 11/02/2014). Besides this, my host was about to construct a house at the outskirts of Kamsar to leave the company apartment house by the time of his retirement. As the retirement was to happen in 2015 and he did not seem to have savings, he had to calculate with a very tight budget, which barely left money for anything but food and construction work. This became evident a few days after this rather short night under his roof, when he called me because his car was out of fuel and asked me to buy half a liter at the roadside just where his car had stopped (Interview with Worker CBG 11/02/2014).

Impoverishment also often starts with the (sudden) death of active workers. A talk with a 14-year-old school boy beside the company swimming pool in Sangaredi shed some light on this problem. The boy was admitted to the protected recreation center because his deceased father had been working for CBG and used this privilege to practice his English and beg for money from foreigners like me. When I confronted him with my impression that his position in Sangaredi was rather privileged compared to the children on the streets outside the protected area, he explained that his mother received no more than 100,000 GNF (about 15 USD) monthly as a widow’s pension, while his school fees already amounted to 60,000 GNF per month. By losing his father, this boy thus lost a considerable part of his privileges. In a way, he dropped out of the CBG polity.

Besides these temporal limits of the protected space of CBG that are visible to every workers’ family, most people in Kamsar and Sangaredi actually work for CBG with limited or no access to improved social services or fringe benefits. If we take into consideration the global chains of command of the cities, this concerns employees of subcontractors as well as other service providers, traders, administrators and even farmers. Already the contract workers and the employees of subcontractors often live below the poverty line. Many undertake the same work as their regularly employed colleagues but earn only one-quarter or less of their salary. This is aggravated by corrupt deals between the owners of subcontracting firms and their contacts in the company, which often result in a further pillaging of the contract workers' salaries by up to two-thirds (Interview with Trade unionist 17/04/2014). An acquaintance in Kamsar worked as a radio DJ at the local CBG owned radio. Despite his considerable popularity, he was only employed through a subcontractor, which meant that he earned only about 140 USD and was only repaid half of the medical bills of his wife, his three kids and himself (Interview with Journalist 13/04/2014). Contract workers include guardians, cleaning personnel, ward assistants and nurses. Guardians and cleaners at the hospital did not even receive the minimum wage. A guard in front of the "White House" in Kamsar stated that he received 250,000 GNF (about 35 USD) monthly for working half-time (Interview with Security guard 20/02/2014). This was confirmed by another security at the ANAIM hospital, who himself had to make ends meet with a wage of 300,000 GNF (about one Euro per day) for full-time work, which meant four 12-h shifts per week (Interview with Trade unionist A 25/02/2014). Some of his colleagues who worked half-time received a pocket money of 150,000 GNF or about 20 USD (Interview with Security guard 21/02/2014).

The ANAIM hospital employs about 100 core workers who receive housing, food provisions and other fringe benefits (Interview with Trade unionist A 25/02/2014). Owing to budget constraints, another 100 employees undertaking pretty much the same work are contract workers: their contracts are not permanent, they receive much lower wages and they do not have any fringe benefits. For instance, ward assistants receive about 100 USD (700,000 GNF) and nurses between 145 and 220 USD (1–1.5 million GNF) a month (Interview with Hospital director 19/02/2014). This category has grown significantly in number in the recent years, to the detriment of the number of permanent employees (Interview with Hospital director 19/02/2014). Maintenance and security are given out to subcontractors (e.g., to the *Entretien Général du Bâtiment* E.G.B; Trade unionist A 25/02/2014) who mostly pay salaries below the minimum wage. Two legal experts in Guinea confirmed that this constitutes an infraction of national law by CBG (Interview with Consultant 17/03/2014; Legal expert 11/03/2014). Besides the violation of Article 94 of the new mining code that regulates responsibilities of the parent company for its subcontractors,[118] Sandra Nichols also sees a violation of §155, which forces the companies to sign a draft code of conduct in which the responsibility for subcontractors is also stipulated (Interview with Consultant 17/03/2014; République de Guinée 2013a).[119] To my surprise, a construction worker reported about similar conditions on a site of the American junior company GAC (see below) in Kamsar. He was paid 250,000 GNF monthly plus a daily allowance of 7,000 GNF, which is also below the minimum wage. In addition, he was promised to be kept after his initial task was finished, which did not happen (Interview with Unemployed 25/02/2014).

3.2.2.1 Electricity, Security, and Relations with the Local Administration

As in the other towns, protests among the population mainly concern water and electricity. According to ANAIM, about 80% of the population "is deprived of an acceptable level of basic public services",[120] as they receive neither sufficient drinking water nor electricity (Diawara 2014: 62; Mayor of Kamsar 17/02/2014). A concerned teacher thus spoke of "total separation" (Interview with Sociologist 22/02/2014) in the city: only the about 50,000 inhabitants of Kamsar cité have always been connected to the grid. In 2010, the Guinean company SMS Kakandé signed a contract with the government that covered the supply of electricity for Kamsar village. With financial assistance from the World Bank, CBG installed a 4 MW generator in need of 150 L of fuel per day. However, as the operator became heavily indebted to CBG—apparently because the population did not pay for the fuel—his contract with the Guinean government was not extended in December 2012 (Interview with Executive CBG 18/02/2014; Mayor of Kamsar 17/02/2014).[121] With the end of this contract, the government "took over" by ordering CBG to provide the necessary means for running the generator (Interview with Executive CBG 18/02/2014). For a couple of months, with longer periods of interruption, *Kamsar village* thereby received electricity for several hours a day (Interview with Mayor of Kamsar 17/02/2014). Alpha Condé thereby temporarily calmed the situation in Kamsar on several occasions, although no sustainable solution has been found to date. Given that the government has not yet refunded CBG, the company has repeatedly threatened to stop the provision of fuel.[122] During my stay, CBG and the sub-prefecture negotiated with French and American companies on a community solar energy project in Kamsar (Interview with Entrepreneur 19/02/2014). The supply of water is subject to similar problems and had been given out to a private company in 2008 (Interview with Mayor of Kamsar 17/02/2014).

The protests for electricity and water regularly end up in pillages and clashes with police, gendarmerie and military, leading to several stand stills of the factory and at least nine deaths since the 1980s.[123] The last of these demonstrations took place on 16 January 2012, because SMS *Kankandé* did not provide electricity to *Kamsar village* as scheduled and intended to raise its prices (Camara 2014). One of the protesters was killed by the police and the police station and the sub-prefecture were pillaged (Camara 2014; Vice mayor of Kamsar 19/02/2014). Both buildings lie at the entrance of the city center. According to the police commissioner of Kamsar, the demonstrators never made it into the city until today. Most demonstrations result in the destruction of police and sub-prefectural buildings, but the infrastructure of CBG had not yet been damaged. When I asked how this was possible, the commissioner replied:

"In such a case, we are ready to sacrifice the life of someone who attacks the CBG! Our mission here is the CBG. We have to do everything to protect the property of the CBG. It's the CBG which pays more than how many million [sic] Guinean civil servants!" (Interview with Police Commissioner 17/02/2014; translation by JK).

Several informants reported on the harsh methods of the military, once it had been sent into dissolve demonstrations (Interview with Mayor of Kamsar 17/02/2014; Vice mayor of Kamsar 19/02/2014). During the last emeutes, when trying to prevent people from entering the sub-prefecture, the deputy mayor of Kamsar was seized and "tortured" (humiliated and repeatedly beaten with sticks) by soldiers (Interview with Vice mayor of Kamsar 19/02/2014). In 2010, one demonstrator died after police and gendarmerie in Kamsar was supported by forces from Boké. Despite his rather biased description of his mission in Kamsar, the police commissioner also complained that CBG neither paid local taxes nor provided sufficient social services. When I asked about a destroyed cabinet next to his office desk, he replied:

"You see, until now we have the consequences. I preferred to leave it like that, because, do you understand? But CBG has not done anything, you see? CBG did not want to do anything for us" (Interview with Police Commissioner 17/02/2014; translation by JK).

The commissioner himself is accommodated by the CBG, receives 40 L of fuel per month and has access to the cheaper foodstuffs of CBG (Interview with Police Commissioner 17/02/2014). Many of his colleagues enjoy similar privileges. In addition to food supplies from the state, about half of the 45 police men in Kamsar profit from the above-mentioned subsidized food provisions from abroad. Furthermore, several other police officers are accommodated by the company (Interview with Police Commissioner 17/02/2014). Similar assistance is extended to the gendarmerie, the marine (Kamsar is a coastal city), customs officers, and the military (Interview with Sous-préfet of Kamsar 12/02/2014). The bonds between mining companies and security forces are thus tighter in Kamsar than in Fria and Débélé. According to the police commissioner, relations between the police and CBG have ameliorated, since the last managing director, Kémoko Touré, had left. Besides trying to cut food supplies for non-workers, Touré had also banished the giving out of fuel to police patrols. The police commissioner of Kamsar also commented on this issue in a relatively straightforward manner:

"You see, when you work for the factories, the expectations are even higher. Sometimes when we want to patrol, for example, we ask for fuel. There is none. That's why I said that there have not been healthy relations" (Interview with Police Commissioner 17/02/2014; translation by JK).

According to the head of human resources at CBG, these continuous requests for fuel constitute a considerable financial burden for the company (Interview with Executive CBG 17/02/2014), not least because there also seems to be a strong lobby from within. A trade union leader stated that senior officials were able to buy cars considerably cheaper via the company and received 100–150 L of fuel per month (Interview with Trade unionist B 25/02/2014). Several informants spoke of millions of USD going into these and other practices, which can rather be called theft or corruption (e.g., Entrepreneur 19/02/2014; Barry 2013b: 7).

Similar arrangements concerning fringe benefits also exist for the local administration. About 60 public servants are accommodated by the CBG and around 200 also receive food supplies. Most of these beneficiaries belong to the higher ranks of the different hierarchies, including the sub-prefect herself (Interview with Sous-préfet of Kamsar 12/02/2014). About one-third of the about 250–300 teachers in Kamsar get food supplies and 20–30 also receive accommodation (Interview with Administrator Kamsar 10/02/2014). Besides the public establishments, this assistance was even extended to two privately-owned schools (Interview with Administrator Kamsar 10/02/2014). Concerning her own situation, the sub-prefect of Kamsar made the following statement:

"I personally, since my nomination, have received no car. Write this down! I haven't received a car, I have not... I have received no car, I have received no fuel bonus, I have received no bonus for the hospital. I did not get boni. Nothing! The CBG has given me nothing! No car, no fuel bonus, no hospital bonus. But it has given me accommodation. That's all. [...] Before, all the sub-prefects who have been nominated here have received cars. Only the sub-prefect of Kamsar, I have received none" (Interview with Sous-préfet of Kamsar 12/02/2014; translation by JK).

By the end of 2013, Kémoko Touré stopped the food supplies for non-workers "because of governance and tax issues" (Interview with Administrator Kamsar 13/04/2014). However, due to extensive opposition—particularly due to the threat of local teachers to expulse all workers' kids from school—his successor reintroduced the supply again in March 2014.[124] A stop of food supplies would have halved the purchasing power of many civil servants (Interview with Sous-préfet of Kamsar 12/02/2014). Many workers did not support the protests of the civil servants, arguing that the public service in other Guinean cities did not receive this form of assistance. One worker also maintained that the level of control at CBG was much higher than in the public administration. In comparison with teachers and administrators, who were able to schedule their working time quite independently, several workers stated that "we are not free, we are slaves" (Interview with Worker CBG 11/02/2014).

The relation between the mayor of Kamsar, Thiani Sylla, and CBG can be described as ambiguous at best. He had himself been employed at CBG from 1973 to 2005, but then seemed to become a problem for the company. According to the head of human resources of CBG, Sylla's destructive force at demonstrations had been so worrying that in 2013, the company paid him a trip to Mecca in the hope for "spiritual appeasement" (Interview with Executive CBG 17/02/2014). In other companies, this practice is normally limited to proper employees. At present, CBK sends six Muslims to Mecca and three Christians to Rome each year

(Interview with Employee CBK 18/03/2014). Until the crisis in 2012, this was also the case in Fria. *La voix de Rusal* reported that Rusal had sent 12 pilgrims to Mecca in 2010 and another 12 in 2011 (Deguenon and Traoré 2011g: 11). The vice mayor of Kamsar confirmed that Sylla had been sent to Mecca by CBG for more than a month in 2013 despite having left the company already 8 years before (Interview with Vice mayor of Kamsar 19/02/2014).

3.2.2.2 Taxes Versus CSR

However, the Hajj does not seem to have made El Hadj Thiani Sylla less refractory. In an interview, he maintained that the relations with CBG were troubled, because the administration neither received any local taxes from the company nor did it sufficiently include the administration into its infrastructural projects (Interview with Mayor of Kamsar 17/02/2014). While other mining towns and cities, such as Dinguiraye, Siguiri, and even Kindia, received considerable sums of local taxes, the administration of Kamsar was still forced to act as a solicitor versus the company (Interview with Mayor of Kamsar 17/02/2014). The only local tax of CBG that figures in the sub-prefectural budget is the so-called taxe sur le chiffre d'affaire (turnover tax). Contrary to its name, this tax comprises a lump sum of 200 million GNF for the whole prefecture of Boké that has not been changed since its introduction in the 1990s. Kamsar receives 50 million GNF (today about 7000 USD) per year, Sangaredi 35 million, or about 5000 USD (Interview with Administrator Sangaredi 28/01/2014; Administrator Kamsar B 13/02/2014).

Since 2010, the administration of Kamsar also produces a 3-year plan of local development (PDLs), which mainly comprises lists of possible infrastructural projects and is elaborated in a complex bottom–up process with the help of the Canadian NGO CECI. This particular project of CECI is mainly funded by Rio Tinto, the French development agency, and the World Bank (CECI 2014). CBG and other donors then choose several projects out of this list (Interview with Consultant 30/01/2014; Administrator Sangaredi 30/01/2014). Sylla described the cooperation between CBG and the administration in local development as "very very difficult", because "we don't speak with one another" (Interview with Mayor of Kamsar 17/02/2014). According to him, CBG does not involve the local authorities in these selection processes and sometimes even pushes for the retroactive inclusion of projects that have not figured on the list before:

"They don't even respect this. If I tell you that they don't respect: in theory [...] they should consult the PDL [...] and then they should consult us. Do you see? We want, we have verified, we have looked into the PDL and we would like to assist you concerning this and that. Do you see? We, we know the priorities. It's our task to tell them: no problem, but accept to give priority to this rather than this" (Interview with Mayor of Kamsar 17/02/2014; translation by JK; Consultant 21/02/2014).

Other civil servants at the sub-prefecture perceive the relation with CBG less negative. For instance, the vice mayor stated that the working climate with the responsible officer of CBG for Projets communautaires is quite good and that meetings take place two or three times per month (Interview with Vice mayor of Kamsar 19/02/2014; Administrator Kamsar B 13/02/2014). In addition, the sub-prefect, Sama Kaba Kourouma, seemed to be much more optimistic than the mayor (Interview with Sous-préfet of Kamsar 12/02/2014). In recent years, CBG has also taken the first steps towards handing over budget responsibility by granting the administration the right to choose the construction firms for the respective infrastructural projects (Interview with Administrator Kamsar B 13/02/2014). Sylla remarked that this half-hearted sign of trust falls back on the administration if the entrepreneurs are not paid in time, which has happened several times (Interview with Mayor of Kamsar 17/02/2014).

3.2.2.3 Comparison of "Public" and "Private" Investments in Infrastructure

In the following section, I will use the relatively good documentation in Kamsar to quantify the relationship between expenditure of CBG "for the public good" and those of the Guinean government and its representatives in the sub-prefecture. From 2011 to 2013, the sub-prefecture received about 60,000 USD from the World Bank and other funds via the *Programme d'appui aux communautés villageoises* (PACV) and registered proper funds (local taxes and royalties apart from CBG) of about 40,000 USD (Interview with Administrator Kamsar B 13/02/2014). Half of these proper funds were spent on administrational wages (Interview with Administrator Kamsar B 13/02/2014). The PACV started in 1999 and aimed at improving the capacities of local authorities by establishing *Plans de Développement Locaux* (PDL). It remains arguably the most needed investment in mining towns such as Kamsar. However, this type of funding was banned from mining towns until recently, justified with the argument that the population already profit from the investments of the mining companies. As a reaction to this lack of capacity building, Rio Tinto started a similar program called *Projet de Renforcement des Capacités des CRD de Boké* in 2006 in Kamsar, Sangaredi, and Kolabui (Diallo et al. 2011: 50–52, 114; Barry 2006a: 33).

Apart from this, Rio Tinto and Alcoa invested about 190,000 USD in 2012 and 2013 for roads and bikeways in Kabata, at the outskirts of Kamsar, which found their way into the budget of Kamsar. The expenditure was connected to a well-founded local development project in cooperation with CECI, which also comprised the agricultural sector and local capacity building. The project was also integrated into the PDL, although it was driven by the economic interests of the companies that are about to start a new refining project in this area. With similar amounts of local tax payments, the sub-prefectural authorities would have doubtlessly done something else.[125] In 2012, the sub-prefectural treasurer registered investments in infrastructural projects by CBG of about 100,000 USD, which were included in the budget but—according to him—did not respect the PDL, despite several efforts of the treasurer (Interview with Administrator Kamsar B 13/02/2014).

The "public" investment of CBG outside the sub-prefectural budget can be estimated between 1 and 3 million USD, depending on definitions and sources. In 2013, CBG stated in a newspaper article that the "assistance for neighboring communities in the form of the basic services, infrastructure realizations, and the creation of employment beyond CBG"[126] amounted to 2 million USD per year (Barry 2013b: 9). This concerns all the sites of CBG and thereby also Sangaredi and Boké. However, many core investments, such as the maintenance of the railway, routes, electricity, etc., are probably not included. The superintendent of the section *Projets communautaires* of CBG maintained that the expenditure for water and electricity for non-workers in Kamsar, Boké, Sangaredi, and other communities amounts to another 2.15 million USD (Interview with Executive CBG 18/02/2014).

Several senior officials of CBG mentioned an annual budget of 500,000 USD for *Projets communautaires* (Interview with Executive CBG 20/02/2014; Executive CBG 18/02/2014). The competent executive even talked about 500,000 USD annually since 1987, which means that the real investment in the 1980s was at least twice that of today (Interview with Executive CBG 18/02/2014). In 2013, a CBG report listed expenses of 335,000 USD for Projets communautaires. Kamsar received 165,000 and thereby about half of the amount of it. The missing 165,000 USD from the above-mentioned sum of 500,000 USD could stem from a change in policies that enforced stronger controls and obliged the accountants to stop payments if the projects did not evolve as provided (Interview with Executive CBG 20/02/2014). These 165,000 USD comprised construction work and projects on education, health, and the environment.[127]

Based on this information, I would suggest that 2 million USD be calculated for "basic services, infrastructure realizations and the creation of employment" (projets communautaires included) and another 2.15 million USD for water and electricity for all areas, where the company operates (see also Moore Stephens 2013: 49). If we suggest that half of this sum goes to Kamsar, this leads to investments in the public good by the CBG of roughly 2 million USD ($4.15/2 = 2.075$). This does not include investments with more direct profitability, such as investments in the workforce, in machinery, and infrastructure (e.g., the passenger train). In 2011, the chamber of mines stated that CBG had renovated 450 workers' houses in the previous year (Chambre des Mines de Guinée 2011: 32). At least half of the workers' houses belong to the state (ANAIM) and many of them are inhabited by public servants (teachers, police men, customs officers, administrators, etc.). Ultimately, these figures are missing stringent definitions to differentiate between "public" and "private" investments. Despite several attempts and even a contract with CBG concerning my research in Kamsar, I did not succeed in obtaining more comprehensive data.[128]

If we also consider the budgets of the other three companies that have already started to invest in Sangaredi—GAC, Rusal, and Henan Chine (see Fig. 1.6)—this figure doubles at least. GAC alone has spent 25 million USD for ecological and social purposes in Guinea since its start of production in 2000. In 2014, the budget for the local communities was 2.5 million USD (Interview with Executive GAC 1/02/2014). This sum is again not solely spent in Kamsar, but also in Conakry and other areas where GAC is active. I therefore propose calculating with 2.5 million USD for the "public expenditures" of GAC, Rusal and Henan Chine in Kamsar.

For 2011 to 2013 and based upon the existing figures from the sub-prefecture, CBG and other sources, the investment in local development at the level of the sub-prefecture Kamsar could thereby be divided into three levels of democratic control:

- First, local taxes of about 40,000 USD, leading to a stable investment budget of about 20,000 USD per year and the state-controlled PACV financed by the World Bank and by other funds, bringing in about 60,000 USD in the three analyzed years. With this, the sub-prefecture of Kamsar can build two classrooms per year.
- Second, investments of new mining projects like that of Alcoa and Rio Tinto in Kabata and certain CSR measures (*Projets communautaires*) of CBG that were included into the sub-prefectural budget. These investments of about 100,000 USD per year were mainly made to appease the population in the construction zone. The respect of the companies for the PDL concerning these projects is disputed.[129] It is clear that the administration only included their projects after the end of the local consultation process of the PDL and also amended the budget of 2012 to include the construction works (Interview with Administrator Kamsar B 13/02/2014). Statements of administrators and representatives of the companies concerning this issue were sometimes antithetic.
- Third, about 2 million USD fully controlled by CBG[130] and at least another 2.5 million USD spent by the other companies working in Sangaredi(Interview with Executive GAC 1/02/2014)[131].

On this cautious basis, we can state that from 2011 to 2013, less than 1% (40,000 USD divided by 4.5 million USD) of the investments in the public good were fully controlled by the *commune urbaine* (which is the same as the sub-prefecture) of Kamsar. If we consider that (1) other operative zones of CBG (Sangaredi, Boké, etc.) are much smaller and thereby attract less social investment of CBG[132] and (2) many "profitable" investments of CBG are in fact of use to the population, this figure could easily be brought down further. The rest can be considered measures of CSR and thereby a continuation of colonial paternalism, with the important difference that the colonial state has been replaced by transnational companies (backed by internationally influential governments) with quite different rationalities and less coherent interests.

When asked why it does not entrust the local administration with the financial control of its development projects, CBG points to corruption and maintains that its social investment schemes cover more than the sub-prefecture of Kamsar. In most cases, the responsible actors additionally play the usual *jeux d'échelle* game, making other persons on other decision levels responsible (Interview with Administrator Kamsar B 13/02/2014). The long-term outcome of these practices is considerable confusion about the notions of governance and government. While the officials of CBG maintain that "the government does nothing! Nothing at all!" (Interview with Executive CBG 20/02/2014), the Guinean chamber of mines confuses CSR with good governance:

"CSR is neither sponsoring nor philanthropy. It is more than punctual social and environmental actions. We talk about good governance, but also about community involvement" (Chambre des Mines de Guinée 2011: 17; translation by JK).

With the new mining code, CBG is forced to spend a small amount of its current CSR expenditures as a contribution to local development. Apparently, the code itself does not force the companies to hand over these taxes to the local authorities at the prefectural level. At the time of writing, CBG preferred to pay these contributions in the form of direct investments (Interview with Consultant 17/03/2014).

3.2.2.4 Schools and Hospitals

The public schools in Kamsar have the same problems as most other Guinean schools. Due to a lack of teachers, they are too crowded to provide even basic education. On average, classes count 45–50 pupils, although classes of over hundred are not unusual (Interview with Administrator Kamsar 10/02/2014). During a reunion, a teacher stated that she had two classes of 128 pupils per day (Interview with Sociologist 22/02/2014). The whole sub-prefecture of Kamsar counts about 30,000 pupils, with about two-thirds of them being children or protégés of workers (Interview with Administrator Kamsar 10/02/2014). The largest school of Kamsar, the *École filima*, counts more than 5000 pupils (Interview with Administrator Kamsar 10/02/2014). As in other mining towns, this lack of teachers catalyzed the formation of private schools, whose teachers are often less educated and less paid, but whose educational outcomes are, nonetheless, better because the classes are smaller. Many of these private schools are owned by former workers

(Interview with Administrator Kamsar A 13/02/2014; Administrator Kamsar 10/02/2014).

The hospital of CBG, simply called ANAIM by the population, lies at the center of *Kamsar* cité and is without doubt one of the best hospitals in the region (cf. Compagnie des Bauxites de Guinée 1983: 28). Having been treated in the hospital myself during the start of the Ebola epidemic, I would, nonetheless, add that it is far from reaching standards of the OECD world. Due to limited funding, evacuations occur frequently. Until recently, most patients were brought to hospitals in Pittsburgh (the headquarters of Alcoa), Brussels and Paris, whereas now they are also transferred to Morocco and even to Conakry. Since the takeover of Condé, two university hospitals have attracted more and more investors and have thereby outpaced the hospital CBG in several respects (Interview with Hospital director 19/02/2014). The director of the hospital insisted that the hospital is not "privé privé", meaning that emergency patients who cannot pay are not turned away:

"That's why I said that the operating system is private. The way of managing things. But at the same time, it is not "private private", because, well, the neighboring population, if they come in situations of urgency, we do not have the right to turn them away. We are obliged to take them. The private facilities are not forced to take them" (Interview with Hospital director 19/02/2014; translation by JK).

This results in about 50% of the receipts not being collected because the patients cannot pay (Interview with Hospital director 19/02/2014; Trade unionist A 25/02/2014). This tendency was reinforced by a decree of Condé that made all births in Guinea free of charge, also in more complicated cases like caesarean sections. Births at "ANAIM" quadrupled after this decree, forcing the personnel in several cases to let the women give birth on the floor because there were not enough beds (Interview with Hospital director 19/02/2014; cf. Doctor Kassappo 19/02/2014). In this case as well as generally, no official document forces the hospital to treat patients who cannot pay, but the company both seems to assume responsibility and to fear the consequences if it would not:

"If we let them die, the population will react. There will be a movement, there will be riots! That's neither safe for the company. Thus, the company […] cannot close its eyes" (Interview with Hospital director 19/02/2014; translation by JK).

The hospital belongs to the state from "a structural point of view, [...] but it operates, let's say, autonomously [...], financed by CBG" in the form of "subventions" (Interview with Hospital director 19/02/2014). As in the case of the other facilities of ANAIM, the state is thereby the owner but does not contribute to its functioning. The employees of the hospital are paid by CBG and are "défonctionnalisés", which means that they do not enjoy the status of civil servants. However, the financial management has been autonomous until 2012. With the fixed "subvention" of CBG, the director of the hospital had to guarantee for its functioning. Since 2012, the situation is even more complicated as ANAIM—and thereby the mining ministry—assumed control of the management (Interview with Executive ANAIM 13/04/2014; Executive CBG 20/02/2014).

To reduce the influx of non-workers, CBG contributes to the construction of a *Hôpital communautaire* (Interview with Sous-préfet of Kamsar 12/02/2014; Hospital director 19/02/2014) and supports other public health centers by building new tracts, giving out material, taking back medicinal waste and even by clearing the bog holes (Interview with Doctor Kassappo 19/02/2014). Medicaments of the health centers are kept in the morgue of the CBG hospital and are carried to the centers every day (Interview with Doctor Kassappo 19/02/2014). The health center in Kassappo—one of the two functioning public *Centres de santé* in Kamsar—was thus doubled in size and received a generator in May 2012, as a measure against a cholera epidemic to "keep the vaccines cool [and] to do infusions day and night".[133] Unfortunately, 2 years later, the generator was inoperable again and the center was without electricity as before, because there was not enough money to buy fuel. As the health center delivers more than 100 babies per month, this leads to serious problems (Interview with Doctor Kassappo 19/02/2014). If patients cannot be treated in Kassappo, they are evacuated to the hospital of CBG. According to a doctor at Kassappo, the patients are then obliged to pay: if they cannot, "we give them first aid and then we let them go. This happens like that. [But] we take them every time, we tell them to search for money" (Interview with Doctor Kassappo 19/02/2014; translation by JK).

The health center in Kassappo has a budget of about 10,000 USD per year (Interview with Doctor Kassappo 19/02/2014), compared to 3.5 million USD for the hospital ANAIM (Interview with Executive CBG 20/02/2014). With salaries between 10,000 and 13,000 USD, each of the about five expatriate specialists receives more within a month than the whole health center of Kassappo gets in a year. Their salaries together sum up to about half of the hospital expenses on labor. The personnel also comprises a dozen of Guinean doctors who have obtained their diploma in France (Interview with Hospital director 19/02/2014). The head physician at the second public health center in Kamsar earns 1.2 million GNF (about 175 USD), less than the contract nurses at ANAIM. Contract workers in the public health centers earn 30–50 USD per month (Interview with Doctor Kamsar 19/02/2014). The public medical personnel that still receives food supply from CBG probably has a higher income from these subsidies than from their proper wages (cf. Doctor Kassappo 19/02/2014).

3.2.2.5 Environment

The impact of the factory on the neighboring population and the workers led to various health problems, i.e., permanent irritation of the eyes (Interview with Worker CBG 11/02/2014), chronical sinusitis, chest pain (all of this is related to dust; Executive ANAIM 17/02/2014), as well as the proliferation of sexually transmitted diseases, such as AIDS.[134] In general, the high population density in connection with low sanitary standards and the shortage of water facilitates the spread of epidemics, such as cholera. The deforestation of the area strongly reduced the soil fertility and changed the microclimate of Kamsar, rendering subsistence as well as commercial farming increasingly difficult (Interview with Executive ANAIM 17/02/2014; Soumah 2008: 117). This also concerns fishery, which became less profitable due to the reduction of the mangroves and the pollution of the sea (Diallo et al. 2011: 68; Executive ANAIM 17/02/2014). Besides this, the city regularly registers accidents with trains, cars, and other machines related to the mining activity (Interview with Sous-préfet of Kamsar 12/02/2014). According to the mayor of Kamsar, the city has not yet profited from environmental impact studies, although this is about to change with the new projects of Alcoa and Rio Tinto (Interview with Mayor of Kamsar 17/02/2014).

In 1992, CBG started to construct a waste deposal in Bendougou in the outskirts of Kamsar. Already 1 year later, the neighboring population began to protest, apparently because it had not been consulted before the start of constructions. When I arrived in Kamsar in April 2014, the inhabitants of Bendougou had ended a half a year roadblock at the entrance road. They demanded compensation for the contamination of wells, the proliferation of insects, and health problems and claimed that they had been promised a connection to the national grid, well drillings, a hospital ward, and better roads (Interview with Executive ANAIM 17/02/2014). During the negotiations, the populations apparently wanted to talk directly to representatives of CBG and refused to talk to regional authorities (Interview with Executive ANAIM 17/02/2014).

3.2.2.6 New Projects

The CBG plans to double its production until 2028 and its major shareholders—Alcoa and Rio Tinto—have begun construction work for a new alumina refinery in Kabata, north of Kamsar. As we have seen, many such plans have been made throughout the history of Guinea, and in many cases, construction work had been started, but only one project—the factory of Fria—actually began to refine bauxite. Besides CBG, at least four other similar projects were announced in the 2000s in Kamsar—by GAC, AMC, Henan Chine, and Rusal's COBAD, the Compagnie de Bauxite de *Dian-Dian* (see Fig. 1.6; Leno 2014a: 2; Administrator Sangaredi 28/01/2014; Diallo 2013a). If all of these companies fulfilled their promises, world bauxite prices would plummet (Interview with Executive GAC 1/02/2014; Sociologist 22/02/2014). During my stay, Henan Chine and Rusal were still about to conduct impact studies. The International Mining Development is a joint venture of four Chinese state-owned companies, namely, China Henan International Cooperation Group (41%), Yongcheng Coal and Electricity Holding Group (51%), Henan Provincial State Owned Assets Operation Company (4%), and Henan Zhonglian Mining (4%). Similar to their competitors, the companies aim at constructing an alumina refinery (AidData 2015; Administrator Sangaredi 28/01/2014). GAC had already invoked new dreams of industrialization among thousands of Guineans and had also displaced more than 500 people by the time of my arrival (Interview with Executive GAC 1/02/2014). The representative of government relations of GAC, Haskell S. Ward, spoke of GAC's entry as a "historic turning point for the country" (Committee on foreign affairs 2007: 45) and mining minister Ahmed Tidjane Souaré stated that the longstanding "ardent desire of all Guineans [...] to have alumina and aluminum factories now was to become reality (Diallo 2005b: 9). However, since the company was taken over by Mubadala and Dubai Aluminium (DUBAL) in 2012, these plans have been abandoned—despite a contract with the government in April 2014 that stipulated the transformation of bauxite

(Guinéeinformation 2014). Mubadala announced that it would first "test" the Guinean bauxite in its own refineries in the United Arab Emirates (Interview with Sociologist GAC 1/02/2014; Executive GAC 1/02/2014). Shortly before signing the contract with Mubadala, the government had also concluded a bilateral investment treaty (BIT) with Abu Dhabi that went along with the renovation of a public hospital in Conakry, funded by the latter (Bah 2014a: 3).

In the case of CBG, even critical spirits, such as the current mayor of Kamsar, believe that this time is different, mainly because political prospects were promising at the time of the interview (Interview with Mayor of Kamsar 17/02/2014). Even if the plans of Alcoa and Rio Tinto will not be implemented, it is very likely that the population will profit, because the augmentation of production forces CBG to implement the new mining code and new quality standards. The company currently runs an extensive impact study that will particularly bring to the light the problems of people in the countryside who often have neither the means nor the sufficient know-how to insist upon their rights.[135]

3.2.3 Sangaredi

As the other bauxite towns, Sangaredi can be divided into a rather protected realm of workers' districts and a disadvantaged larger population around and within the town, which is excluded from the company's services, but has to endure the negative consequences of the mining activities. Only in 2010, this led to violent clashes and riots that left at least eight inhabitants dead and 15 injured. As I had already been invited by CBG to visit the mining facilities in 2012 and had even been accommodated for free by the company during my stay, I planned to start my second field research in Sangaredi in January 2014 at the outskirts of the city, or at least outside the company realm. However, this proved quite tricky. Through contacts in Conakry, I was allowed to live with a workers' family in Sangaredi, which I could hardly refuse, and from there onwards, all my new contacts came from inside the "walls" of the company. Furthermore, everybody advised me to see the director of the company—and not the local authorities—before starting my work.

Sangaredi is a sub-prefecture with 70–80,000[136] inhabitants. In 2011, CBG registered 1133 permanent workers (core workforce and contract workers) and 100 "seasonal contractants" in Sangaredi (Compagnie des Bauxites de Guinée 2012). The rest of the population lives mainly from agriculture, trade, and to a certain degree also transhumance (Leno 2014a: 4). CBG maintains that before bauxite mining, “the Sangaredi plateau lay in wilderness” and that the “area was virtually unpopulated” and inaccessible (Compagnie des Bauxites de Guinée 1983: 13). Several elders of the villages around the town did not share this view and also the buildings and fruit trees of villages, such as Hamdalay, seem to be older than 40 years. Transhumance certainly existed already before bauxite mining. The town itself was “literally built from the ground by CBG” (ibid: 21). This included the mining headquarters, 489 workers’ houses (Compagnie des Bauxites de Guinée 2012), a mosque, a cultural center (with a movie theater, a library, a performance hall, etc.), a railway station for passengers (Compagnie des Bauxites de Guinée 1983: 21), headquarters of police and gendarmerie (Officer Gendarmerie 27/01/2014), and many other buildings. In addition, the town possesses football, handball, volleyball and basketball grounds, and athletic tracks (Leno 2014a: 6). During my stay, I also assisted a basketball game and several football matches between Guinean schools and clubs.

Until the early 1990s, the exploitation of the bauxite reserves in the sub-prefecture was limited to the so-called Sangaredi plateau with particularly high ore grades. Since 1992, CBG also exploits Bidikoum, since 1997 also Silidara and since 2007 also the reserves of N’Dangara (République de Guinée 2013c: 15). The ores of the different plateaus are mixed to obtain a stable alumina content for the exports (Interview with Executive CBG 27/01/2014).

3.2.3.1 Services

Public and social services in Sangaredi are organized similarly as in the other bauxite towns. CBG maintains the infrastructure of the town, provides services, such as waste management, fire service, and sewerage, and is also responsible for electricity and water (Interview with Engineer CBG 6/02/2014). This includes a passenger train, a very popular, and by far the cheapest way to travel to Kamsar and to several other destinations on the way. Until 1995, this railroad was even more important as there was no asphalted route between Sangaredi and Boké (Leno 2014a: 5).

The whole population in Sangaredi-Centre (about 68,000 inhabitants; Administrator Sangaredi 28/01/2014) receives electricity and water for free, no matter if employed by CBG or not. This takes place based upon “bon voisinage” (good neighborliness). The CBG is not forced by any contract to do so. The same applies for virtually all other activities (Interview with Administrator Sangaredi 28/01/2014). As in the case of Fria, the power generators and the water treatment are directly connected to the mining headquarters (Interview with Administrator Sangaredi 28/01/2014). A stop of production would also mean the end of power and water. This dependence was slightly reduced by the construction of several well drillings by the *Service national des points d'eau* (SENAP). CBG and GAC have also constructed well drillings in Sangaredi, but in many cases, these wells become dysfunctional a few years after their installation and are not subsequently repaired. This situation prevails despite the formation of the so-called *Comités de gestion des forages*, local groups responsible for maintaining the well drillings by locally raising money.[137] Both water and power are sources of conflict in Sangaredi, due to periodic power cuts and because only parts of the town profit from the services (Interview with Consultant 30/01/2014).

The private health sector in Sangaredi is less dominant than the one of Fria and Kamsar. The town has a *Centre de santé amelioré* (CSA) and nine health posts (*Postes de santé*), all of them being state-owned. The "hospital" of CBG has only the status of a clinic (*dispensaire*), which—according to the ministry of health—is comparable to the size of a health post (Interview with Administrator Conakry 26/03/2014) and has no maternity ward (Interview with Hospital director 19/02/2014).[138] In fact, however, it still seems to be at least as important to the population as the CSA. The latter registered 15,235 consultations and 372 hospitalizations in the second half of 2013 and disposed over a budget of 99,030,058 GNF (about 14,000 USD) for the same period (Interview with Doctors 29/01/2014). These funds are mainly controlled by the ministry of health. Only some maintenance personnel are additionally paid from local taxes (Interview with Administrator Sangaredi 30/01/2014). The private clinic registered 16,000 consultations and 1400 hospitalizations throughout 2012 (Compagnie des Bauxites de Guinée 2012). The significantly higher share of hospitalizations is due to a better equipment (e.g., a blood bank) as well as better trained and much better paid employees at the clinic. The relatively low number of consultations results from the fact that the clinic does not deliver babies and is not responsible for the treatment of pregnant women. Even pregnant (wives of) CBG workers are treated at the CSA (Interview with Doctors 29/01/2014). The clinic frequently accepts urgent cases from the public health facilities, but is, otherwise, quite strictly limited to the treatment of CBG workers (Interview with Administrator Sangaredi 28/01/2014).

According to the CSA's procedures, patients who cannot be treated have to be evacuated to Boké. However, the patients often demand to be delivered to the clinic in Sangaredi (Interview with Doctors 29/01/2014). These services for non-workers are not free of charge as is de facto the case in Kamsar (Interview with Administrator Sangaredi 28/01/2014). Since the public health sector is neither substantially subsidized by the mining companies (Interview with Doctors 29/01/2014), this leads to a *relatively* independent public health sector in Sangaredi. A shutdown of the clinic in Sangaredi would lead to a deterioration of public health, although according to several doctors at the public health center, the consequences would be manageable:

"The disappearance of the clinic CBG, well, won't affect the population's health that much. You know? The clinic of the CBG does not carry out preventive measures and it's the preventive measures that lead to the improvement of public health. [...] It's our health center which is in charge of that" (Interview with Doctors 29/01/2014; translation by JK; Administrator Sangaredi 28/01/2014).

The educational system in Sangaredi can also be compared with the other mining towns. While the state runs 30 primary and three secondary schools, non-governmental actors possess 17 primary and six secondary schools (Interview with Administrator Sangaredi 29/01/2014). The three public secondary schools host almost 6000 pupils, while the private ones only account for about 2000. The largest public secondary school alone, the *Lycée Général Lansana Conté*, provides rooms for more than 2800 children (Interview with Administrator Sangaredi 29/01/2014). According to the local authorities, most of the private schools are owned by senior officials of CBG (Interview with Administrator Sangaredi 29/01/2014; Administrator Sangaredi 3/02/2014). As in the other mining towns, private operators pay significantly lower salaries (50–120 USD as compared to 150–400 USD in public schools) and collect much higher school fees (70–110 USD per year compared to about 1.2 USD in public schools). This leads to significantly smaller classes (30–40 pupils per class as compared to 100–120 in public schools) and thereby to better learning conditions, despite the personnel being less qualified (Interview with Administrator Sangaredi 29/01/2014). This situation is complicated by the fact that public teachers sometimes also work in private schools. However, these practices seem to be less prominent than in Fria (Interview with Administrator Sangaredi 29/01/2014).

As in Kamsar, the CBG ran a school for the children of expatriates in Sangaredi in the 1970s, but due to the reduction of their numbers, this school has been closed. Today, the activities of the mining companies in education are mainly confined to the construction of schools.[139] However, according to the local authorities, the main educational problem is a lack of teachers. Sangaredi counts many more classrooms than teachers (Interview with Administrator Sangaredi 29/01/2014). Being indirectly financed by CBG, the private schools—both the starting capital and the school fees mainly come from workers' families—contribute heavily to education in the sub-prefecture. Their closure would mean a further overburdening of the public schools (Interview with Administrator Sangaredi 29/01/2014). Compared to the intellectual spirit in Fria, leisure activities in Sangaredi seem to stronger concentrate on sports. One indication of this difference of "company culture" that influenced the whole local area is the availability of books and journals. Despite an existing library and even a literary club (Leno 2014a: 6), books are literally in short supply in Sangaredi. During my stay, I found neither newspapers nor books in the local markets.

3.2.3.2 CSR, PDLs, and Taxes

In Sangaredi, the difference between voluntary projects of the companies and public infrastructure projects was, at the time of writing, particularly vague. In 2014, two "projets sociaux" of CBG in Sangaredi entered the budget of the urban community: the construction of a warehouse in the district Minel and the construction of 13 latrines for a state-owned school. However, CBG, nonetheless, retained the financial aspect of the process and only paid the construction workers after they had already finished half of their work. This often lead to conflicts with the community, which established the contacts with the construction contractors and then unwillingly had to act as a mediator between company and constructor (Interview with Administrator Sangaredi 30/01/2014). In many other social projects of CBG, the company apparently takes into account the PDL, but does not insert the projects into the budget of the local administration. When I asked the *secrétaire général* of the *commune urbaine* whether the local authorities are also involved in these cases, he confirmed, but this cooperation seems to be quite limited:

"Yes they do that. They tell us, they are about to construct infrastructure at this place, do you understand? Currently, that's their approach. Recently, since 2010, they also draw on our development plan [to see] what they can realize" (Interview with Administrator Sangaredi 30/01/2014; translation by JK).

Besides the social projects of CBG that enter the communal budget or not, the general secretary also mentioned the turnover tax as the only tax that the local community currently receives from CBG. It amounts to 35 million GNF, today about 5000 USD, and has not been changed since 1992. However, even for this small tax, the budget authority of the communality seems to be limited. The *secrétaire général* complained that they had not received the TCA (Taxe sur le chiffre d'affaires) for 3 years and was unable to give reasons for this. Only recently, CBG "suggested" building a warehouse with the accumulated TCA of 2011–2014 (Interview with Administrator Sangaredi 30/01/2014). The mayor of Sangaredi maintained that the constructor had already finished this warehouse but still waited for his remuneration from CBG. This means that these "taxes" remain as well under the control of CBG (Interview with Mayor of Sangaredi 30/01/2014). The only other mining tax mentioned by the *secrétaire général* was a surface tax based upon the new mining code, which is paid by Henan Chine and amounted to 226 million GNF in 2012, today about 30–35,000 USD (Interview with Administrator Sangaredi 30/01/2014; Consultant 30/01/2014). In the same year, the *Commune urbaine* (CU) collected about 290 million GNF of "local revenues" (*recettes locales*) and received a PACV assistance of about 380 million GNF that was itself again linked to predefined projects (Interview with Administrator Sangaredi 30/01/2014). The local revenues are paid by small businesses. During my stay, the administration implemented a tax for motor bike taxis, which promptly led a noisy strike of the local taxi union in front of the sub-prefecture (Interview with Consultant 30/01/2014; Administrator Sangaredi 30/01/2014). These efforts of the CU are comprehensible as the local taxes seem to be the only ones that are really controlled by the local administration.[140] However, if we take into consideration that most of these local taxes are paid by inhabitants with incomes below the Guinean minimum wage (about 60 USD per month), it is, nonetheless, important to note that the contribution of CBG to the local budget is virtually non-existent.

A final category of CBG development projects neither enters budgets nor uses the PDL, but could be called ad hoc projects in case of conflicts with the local population. One such project was undertaken in the village of Boundouwouande and led to considerable discontent among both the concerned population and the local authorities (see below).

Rich Workers and Poor Administrators

The living standard of core workers' families in Sangaredi is surely among the highest in Guinea, albeit still very low compared to living conditions in the rest of the transnational production network of the shareholding companies. With the takeover of Fria by Rusal, salaries in Sangaredi and Kamsar became the highest in the Guinea bauxite sector. According to a trade union representative, only gold miners in Siguiri receive more (Interview with Trade unionist 27/01/2014). This creates a high economic impact of workers' salaries for the town and the surrounding villages. Workers have important assets in education, construction, agriculture, animal husbandry, commerce, etc. (Interview with Administrator Sangaredi 3/02/2014) and provide the demand for most other businesses. Due to their unique status, the workers' households are, nonetheless, much larger than normal ones, providing shelter also for distant relatives and often also for *protégés* from poor families (Interview with Administrator Sangaredi 3/02/2014; Administrator Sangaredi 28/01/2014). In the workers' household, where I was allowed to live for several weeks, both parents were employed at CBG. Despite their relatively high income, they had only three biological children who went to private schools, were able to spent weekends in Kamsar, and received additional education, such as IT training (the family does not yet have a computer) and English courses. The family accommodated, nonetheless, another six persons, aged between 7 and 25, who had received only primary education or no education and did most of the household work, such as cooking, shopping, cleaning, and the carrying of water. The house had been constructed by CBG and then bought by the family. It is connected to the water supply network, although the pressure is too low for the water taps to be functional. The comparatively high living standard thus mostly stems from a decent income and stable electricity. However, it is important to note that even in a privileged household like this, running water does not exist. Instead of a shower in the Western sense of the word, the family members still use water buckets carried to the compound from neighboring households (Interview with Employee CBG 7/02/2014).

As has already been shown for the case of Kamsar, the living conditions of contract workers and employees of subcontractors are far below the core workers' standards. One of the *protégés* of the family, where I lived had started to work at a construction site of CBG a few days before I came to live with the family. Like many of his colleagues, he quit work a few days later due to inadequate pay. He had received 25,000 GNF or about 3.5 USD per day (Interview with Unemployed 26/01/2014). This was confirmed by a Gambian foreman at the site, who earned 50,000 GNF a day (Interview with Builder 7/02/2014; Administrator Sangaredi 30/01/2014).

The mayor of Sangaredi is a retired teacher and receives 150,000 GNF (about 20 USD) as an allowance for his political function (Interview with Mayor of Sangaredi 30/01/2014). Like many other mayors, he was not elected, but nominated by the government after many local authorities had been put out of office on corruption charges by the new government in 2010 (Interview with Consultant 30/01/2014; Administrator Sangaredi 3/02/2014). According to the activist Moktar Diallo, the government would have been forced by law to replace these local *Délégations spéciales* by democratically elected authorities long ago (Interview with Activist 2/03/2014). The local administrators receive salaries far below the wages of equally qualified employees of CBG and are only rudimentarily equipped to exert their functions (Interview with Administrator Sangaredi 28/01/2014). During my stay, this led again on several occasions to the situation that local authorities were unable to do their job due to a lack of material means. When I once felt obliged to pay a small "assistance" to an administrator, the outcome felt like having triggered a small revolution: suddenly, the administrator found the time to take me to the villages concerned by the mines and even scheduled further consultation processes with the villagers.

As in Kamsar, the company provides free accommodation for a number of civil servants, including the sub-prefect (Interview with Administrator Sangaredi 28/01/2014) and his deputy, the general secretary of the CU Sangaredi, the commander of the gendarmerie, his deputy, and four other subordinates, the police commissioner, and a considerable number of police officers and many teachers (Interview with Administrator Sangaredi 28/01/2014; Administrator Sangaredi 3/02/2014; *Conseiller Militaire* CBG 5/02/2014). In addition, CBG pays boni to police forces (Interview withEmployee CBG 14/02/2012;

Activist 8/02/2012) and de facto employs another six gendarmes and the conseiller militaire who work together with private security forces and also receive accommodation and several other fringe benefits (Officer Gendarmerie 27/01/2014). A meeting with the young *conseiller militaire* of Sangaredi left a particularly chatoyant impression on me. I waited for him in front of his very spartanly equipped office when he arrived with a new Chrysler Touring 300 that he awkwardly parked in front of the entrance. As he is the representative of ANAIM in Sangaredi, I asked him what he knew about this institution. In quite broken French, he answered that ANAIM was the state and the state also sent him. He did not know anything about the functions of ANAIM. He was paid and accommodated by CBG and disposed over another company car, but preferred to come to office with his Chrysler, which he had received from a "brother" (Interview with Conseiller Militaire CBG 5/02/2014).

3.2.3.3 Local Neopatrimonialism

The highly concentrated infrastructural power in Sangaredi often leads to discontent among local politicians, administrators, and the population. Several interviewees complained that the *Commune urbaine* was on the mercy of the director of the mine. An administrator stated that the director would behave like the "owner of everything" (Interview with Administrator Sangaredi 3/02/2014) in town, occupying sport and cultural facilities and constructing new facilities without asking the public administration. Before the current director took over, the company and the *commune urbaine* met three times a year, but the new director stopped this collaboration (Interview with Administrator Sangaredi 3/02/2014). The mayor of Sangaredi, Mamadou Diouldé Bah, maintained that "it is true that we have gained many benefits, but the disadvantages are more numerous and more visible" (Interview with Mayor of Sangaredi 30/01/2014; translation by JK). For him, this mainly concerns non-existent tax payments on the local level—and thereby the fact that the local authorities have only little influence on investments in the "public good"—as well as the fact that 40 years of profitable mining have not led to the construction of roads for the villages around Sangaredi and did not even allow for electricity and water supply throughout the urban community (Interview with Mayor of Sangaredi 30/01/2014). When I confronted a representative of GAC with these grievances, she defended the measures of CBG. Assuring that she knew the mayor very well, she advised me not to take him all too serious, because "sometimes... what he says...you have to insist a bit because it's not always...I mean... he doesn't give the whole picture. [...] For instance, if you talk to the mayor he will always say that GAC has never associated him, but we have never constructed infrastructure without ... approval by the authorities" (Interview with Executive GAC 1/02/2014; translation by JK).

This would also happen, because the schools and health centers constructed by GAC have to be manned and maintained by the local administration afterwards (Interview with Executive GAC 1/02/2014).

3.2.3.4 The Villages Around Sangaredi

The inhabitants of the villages near the mines and around the town face similar problems as their compatriots in Fria: the dust leads to health problems and renders agriculture inviable, the blastings frighten away the game, destroy buildings, and redirect streams and groundwater, and the surface mining reduces space for cattle and agriculture and cuts the routes between the villages and the town (Interview with Administrator Sangaredi 28/01/2014). The chef de service du *développement rural* of Sangaredi complained that the game also diminished due to wrong restauration of former mining areas (Interview with Administrator Sangaredi 28/01/2014).

Based upon its mining convention with the Guinean state, CBG has reforested 1130 ha of former mining pits until today (Interview with Executive CBG 10/02/2014). Most of these areas are now monocultures, comprising either acacia or cashew nut trees. Animals, such as antelopes, chimpanzees, several other apes, and various birds, did not come back to those areas (Leno 2014a: 2). Interviews with CBG executives showed that this problem has not been tackled to date. One representative argued that restauration is insufficient:

"We, we are improving things. That's what sustainable development means! Sustainable development depends on three factors: the economy, the social situation and the environment. And these three are connected by what we call good governance, which is a transversal value. [...] We have to upgrade. And if you improve you have to take care of ecological, economic and social realities. And in this respect, we have created what we call the synergy with the communities" (Interview with Executive CBG 23/02/2014; translation by JK).

This elaborated discourse of sustainability does not (yet) match the actual reality on the ground. As the communities near the mines are in need for additional income, the company mostly planted cashews. In other instances, the company asked the local population to bring along seeds that were sown after the restoration of the topsoil (Interview with Executive CBG 23/02/2014). Until today, it remains unclear who has the right to harvest these cashew nuts. The chef de *service du développement rural* of Sangaredi remarked that those who think that the nuts are ripe collect them and insisted that this is in fact theft (Interview with Administrator Sangaredi 28/01/2014). While CBG does not seem to feel concerned, this leads to conflicts between the neighboring villages and with other commercial producers in the prefecture of Boké who complain about the low price and the low quality of these cashews (Officer Gendarmerie 27/01/2014; Administrator Sangaredi 28/01/2014). Many of the cashews are exported to China and India. During my stay, my host family in Sangaredi was about to buy a storeroom exclusively for cashew nuts at a newly constructed market. The cashews were to be exported to India with the help of a "cousin" (Interview with Employee CBG 7/02/2014). When asked why the plantations have not yet been given officially to the population, a CBG representative answered that the company had not yet finished mining and thus could not hand over the plantations for security reasons. Once the population was officially given the right to harvest, this would mean that the company could not come back mining again (Interview with Executive CBG 23/02/2014). In the meantime, the harvest and its export are mainly controlled by informal networks of CBG workers.

Relocations and Compensations

Compensations for inflicted damages almost exclusively depend on written complaints by the concerned population. Accordingly, given that almost none of the villagers speaks and writes sufficient French, they are rarely paid. Only in case of official relocations do the concerned villagers normally receive compensation, although the living conditions worsened for the majority of the relocated population in these cases. The responsible CBG executive in Sangaredi reported about relocations in Loumbayodo, Ndanta Fonye, and Bidikoum. As will be shown, all of these relocations led to serious problems for the villagers. Loumbayodo and Ndanta Fonye will be treated in some detail below. The executive stated that the relocation of Bidikoum did not work out because certain brokers wanted to "eat the money of the poor" (*manger l'argent des pauvres*) and added that there was a "management anarchy" (*anarchie de gestion des affaires*) in connection with compensations and relocations (Interview with Executive CBG 10/02/2014). As these cases are treated and controlled by a few persons from CBG and the local administration, it is highly probable that many conflicts and even relocations never became known. For instance, a subcontractor of CBG that I had met by chance when I was hitchhiking from the mine to the town reported on an ongoing conflict in a village called "Bora" which nobody else had mentioned during my stay. According to this informant, the village is situated only 400 m from the mines and had blocked blastings a few days before the talk with the subcontractor (Interview with Employee subcontractor 6/02/2014).

Despite several attempts, I was neither able to get hold of maps that indicated the property of the rural population nor of any impact studies of CBG connected to relocations and compensations. The responsible CBG executive in Sangaredi mentioned studies on Bidikoum, Silidara, and Boundouwouande, but none of the responsible executives of CBG possessed copies of them.[141] However, with the new mining code and due to other international engagements of Alcoa and Rio Tinto (in particular, the "IFC Performance Standards on Social and Environmental Sustainability"), this situation seems to be changing now. Already in 2012, CBG started an extensive impact study with the Canadian consultancy agency "EEM Gestion Durable" for the new mining areas. This agency coordinates a large group of researchers, working on social, economic, and environmental topics (Interview with Executive CBG 23/02/2014; Executive CBG 10/02/2014). In Sangaredi, I met a French researcher of the consulting team INSUCO who solely focused on the preservation of the traditional places of worship (Interview with Anthropologist 4/02/2014). This new approach is also catalyzed by new investors, such as GAC and Henan Chine. For instance, one of the local administrators of Sangaredi had conducted a small social impact study for this Chinese firm only a few months before we met (Interview with Administrator Sangaredi 28/01/2014). However, the current impact of mining on the local population will, nonetheless, remain deplorable until these studies lead to projects and these projects are actually carried out.

Compensations for Destructions

CBG only compensates for the destruction of private property if the victims file a written plaint at the sub-prefecture. This is a high hurdle for the mostly illiterate farmers and renders such plaints rather rare (Interview with Administrator Sangaredi 3/02/2014). A technician at the service environment of CBG only knew of one case of compensation in 2012 and none in 2013 (Interview with Engineer CBG 6/02/2014). The superintendent of the section Projets communautaires of CBG registered 18 plaints for all areas, where CBG has operated since 2010 (Interview with Executive CBG 18/02/2014). These plaints are the basis for the convocation of a "Commission de constat", comprising representatives of CBG, the sub-prefecture, and authorities from the district level who evaluate the destructions. All the participants of the *Commission de constat* receive 200,000 GNF (about 30 USD) from CBG to be able to do their work. This is crucial for the mobility of the involved locals. The *chef de service du développement rural* of Sangaredi then applies fixed compensation standards to determine the exact sum that has to be paid which he sends again to CBG (Interview with Administrator Sangaredi 3/02/2014; Leno 2014b). After approval by the company, he issues a check that is delivered to the victims during a public ceremony (Interview with Administrator Sangaredi 3/02/2014). The payment standards comprise fixed sums for every sort of plant that has not been changed since 1987 and are only to be adjusted to inflation at a rate of 5% each year (Interview with Administrator Sangaredi 30/01/2014). The existing figures on inflation since 1987 rather suggest an inflation rate of 15% per year (IMF 2014).

I was able to follow the actual process of such compensations at the example of a farmer in a remote village who was about to lose 40 cashew nut trees due to a new mining route. The documents of the responsible civil servant suggested a price of 3000 GNF or 0.33 USD per plant in 1987. To calculate the current price, he did not take into consideration the compounded interest of 5% per year and came to a sum of 7000 GNF or about 1 USD per plant. If he had taken the compounded interest into consideration, the initial amount would have more than tripled to 10,667 GNF in 2013 or to quite accurately 150% of the proposed 7000 GNF. If the regulations that the civil servant uses took into consideration the real inflation of about 15% per year, this sum would attain 60,000 GNF per plant or 8.6 times the proposed compensation: for all trees, this would have amounted to 343 USD rather than 40 USD. Eventually, the civil servant also showed me the compensation schemes of the US company GAC, which had just began to develop its mine in Sangaredi and used World Bank standards. In GAC territory, the farmer would have received 80,000 GNF per plant or about 460 USD for his 40 trees (Interview with Farmer 5/02/2014; Administrator Sangaredi 28/01/2014).

These figures show the arbitrariness and the low transparency of existing compensation schemes. The concerned farmer neither knew how to calculate compounded interest rates nor did he speak sufficient French to understand the regulations. Accordingly, what should be simple payouts based upon publicly available schemes become opaque negotiations that leave ample room for petty corruption. According to the farmer, colleagues of the responsible administrator and CBG employees directly approached him after he had started complaining. They started with an "offer" of 5000 GNF per plant and doubled to 10,000 GNF when he threatened to block the routes. As he continued to refuse, he was even offered a "small contract" for 2–4 months of work. The farmer refused again, and at this point, he already had many inhabitants of his village behind him, who threatened to stop the machines unless the farmer received his compensation. An administrator explained that these confrontations have become frequent in recent years:

"The problem is that they don't even care to ask the people: I am coming. They come and they scrape. That's why the population currently takes action: when a machine comes the population comes out and blocks the machine. They say: you don't go further! In theory, they should not go further without having done an impact study with the community. [...] You sign the papers. Then you open up" (Interview with Administrator Sangaredi 28/01/2014; translation by JK).

I was then able to follow further discussions between the farmer and the responsible administrator. The latter explained that the farmer's land had to be registered for a sum of 150,000 GNF before he could demand the compensation and he would have to pay another 200,000 GNF for the damage analysis, which would not be refunded by CBG if the analysis was negative. This already amounts to about 50 USD. The compensation "proposal" of the administrator only amounted to 40 USD. Then, it turned out that a registration of the cashews was not possible because the farmer did not own enough land—only about 2 ha instead of a minimum of 5 ha (Interview with Farmer 5/02/2014; Administrator Sangaredi 28/01/2014). This means that ultimately the farmer had neither any viable claims nor the necessary know-how or contacts to enforce them. He could either hope for benevolent local administrators and CBG executives or continue with his threats to block the road. This situation is paradigmatic for the Guinean countryside. Besides a lack of know-how, the concerned population also lacks the legal basis to face the mining companies and their intermediaries: whereas land titles are mostly limited to urban areas, customary law is either ignored or interpreted to the detriment of the local population (Interview with Legal expert 11/03/2014; Legal expert 13/03/2014).[142] In the following, I will report on the visits of four concerned villages in Sangaredi.

Hamdalay

Hamdalay, situated on the only road that leads to Sangaredi, about 10 km before Sangaredi-Centre, is said to be the oldest village in Sangaredi and probably one of the few that already existed before the construction of the mining town. Due to its location, it is one of the first settlements that became concerned by mining. Today, it is also one of the villages concerned by extensions projects of CBG. I arrived in Hamdalay by motorbike with the village teacher who lives in town. Consisting almost exclusively of round huts made out of clay and shaded by large fruit trees, the village made an almost idyllic impression to me. After mounting and descending several pebbled paths that connect the houses, we met the elder of the village working on his manual sewing machine under a mango tree. He decried vanishing rivers and game, dried up wells, and ever-shrinking space for agriculture and cattle raising. Only in 2005, the village has received a school with three classrooms in compensation for the inflicted damages. According to the inhabitants, this was the first investment of the mining companies since the start of mining and even the equipment of the school had been paid by the villagers themselves (Interview with Elders 31/01/2014). Due to this conflictive history, relocation plans frightened the villagers (Interview with Teacher 31/01/2014; Barry 2013b). They demanded meetings with the executives in charge and sufficient time to take decisions based upon CBG's proposals (Interview with Elders 31/01/2014). After this interview, I met the chief of the village. According to him, only two out of 500 villagers work for CBG, one of whom was himself. He was employed as a watch, although his job seemed related more to mediation rather than guarding:

"[The Watches] are not really there to work for CBG, no, they are between CBG [and the neighboring population], to prevent the people from destroying down there so that CBG can do things [...]. Because they know the youth of the village. [...] They are intermediaries"

(Interview with Elders 31/01/2014; simultaneously translated from Peul into French by the teacher of Hamdalay; then translated into English by JK).

The chief thus occupies a typical broker position as it had already existed in colonial times (cf. Bierschenk and Olivier Sardan 2002). He complained that CBG had only complicated the lives of the community to date and he showed understanding for the enraged "youth" who stole and destroyed company property in response (Interview with Elders 31/01/2014; Employee CBG 31/01/2014). At the same time, he was paid by the companies to limit these infractions. When asked if he was only employed to calm down (apaiser) the population, he answered:

"Of course. It's to calm down. But they [the watches; note by JK] are there to canalize the people so they won't revolt. Every time there are tensions, […] they will calm the youth of the village so that they won't go to spoil things until hearing the truth between them and the CBG" (Interview with Elders 31/01/2014; simultaneously translated from Peul into French by Mr. Bah, teacher at Hamdalay; then translated into English by JK).

Ndanta Fonye

Another concerned village, Ndanta Fonye, is situated a few kilometers northeast of Sangaredi. I went there by motorbike with an administrator, who blamed the company for "caging" the population "like animals" in this area (Interview with Administrator Sangaredi 28/01/2014). A previously existing route had been closed due to mining activities and the remaining routes lead over kilometers of pointed rocks that were very difficult to cross on the backseat of his motorbike. We were welcomed by a dozen of elders and their families in front of the chief's house and were given a meal that consisted exclusively of products from the same village.

The inhabitants of Ndanta Fonye had been relocatedin 1985 from the village Loumbayodo and had been recompensed in cash. They received 50,000–150,000 GNF (today probably about 300–900 USD) for their houses and small amounts for cash crops (e.g., about one USD for a mango tree). Altogether, the village of about 1000–1500 inhabitants received 2,794,000 GNF, today about 15,000 USD (Interviews Reunion in Ndanta Fonye 5/02/2014). According to the interviewees in the village, CBG suggested Ndanta Fonye as the new home for the population and assured that there was no bauxite under the Silidara plateau, where the village was to be constructed. However, in the 1990s, the company changed its mind and also started to destroy the new plantations next to the new village, this time without any compensation (Interviews Reunion in Ndanta Fonye 5/02/2014; cf. République de Guinée 2013c). After the caterpillars had done their job, the company planted cashew nut trees as rehabilitation. The interviewees complained that these trees rendered the soil very dry and did not allow for the fauna that had previously flourished in the area (Interview with Administrator Sangaredi 28/01/2014). In addition, the company never officially assigned the trees to the population. The collection of the cashews is thereby still considered illegal by the company and the local authorities (Interviews Reunion in Ndanta Fonye 5/02/2014). CBG also promised to construct wells, but instead, its bulldozers only dug artificial water holes that filled up with mineral wastes during the rainy season and thereby became undrinkable. Instead, the population has to rely on self-dug wells without water treatment. The interviewees related this situation to the low life expectancy in the village. According to them, the eldest inhabitants of Ndanta Fonye are around 50 years (Interviews Reunion in Ndanta Fonye 5/02/2014). The situation in the old village was considered much better in comparison with Ndanta Fonye, mainly due to the existence of potable water and favorable conditions for agriculture. Until today, the population also cultivates in the remaining area around the old village (Interview with Farmer 5/02/2014).

These and other incidences created longstanding animosities of the population versus CBG. According to the administrator who had taken me there, the population no longer wants to speak to CBG officials, because they have only worsened their lives to date (Interview with Administrator Sangaredi 28/01/2014). Only recently, the inhabitants had stopped the mining machines by sitting on them (Interview with Teacher 31/01/2014). After the interviews, the administrator suggested convening an assembly of all concerned villages, which should subsequently report to the national assembly to put pressure on CBG. The population agreed, although to my knowledge, this has not happened at the time of writing. When I asked the administrator a few days later in his office if he had already contacted other concerned villages, he maintained that this had to be done by the villagers themselves and that neither the administration nor CBG would act before the villagers had filed a written complaint (Interview with Administrator Sangaredi 7/02/2014). However, none of the persons we met in Ndanta Fonye was even able to *speak* French.

Wendou Diawle

The next large resettlement after Loumbayodo in 1985 concerned the inhabitants of Wendou Diawle, North of the Silidara plateau, in 1992. The representatives of CBG seemed to have learned from the case of Loumbayodo and constructed new houses and also new wells for the population (Interview with Administrator Sangaredi 28/01/2014). However, in this case, the population complained that the houses had not been finished and that other promises had been broken. The executives had promised to employ some of the villagers and construct roads, both of which had not been done. As in Ndanta Fonye, the main problem is a lack of potable water. Given that the wells no longer function anymore, the population has to search water in the neighboring villages, about 1.5 km from Wendou Diawle. As in the case of other villages, the population had dug their own small well beside a river, but shortly before my arrival, this well collapsed. In the old village, water was only a problem during 1 month in the dry seasons, whereas the rest of the time, the villagers were even able to fish in the streams nearby. In addition, large areas around the village have been deforested by town dwellers who were entitled to land in this area and—like in Ndanta Fonye—the new location was rapidly encircled by the mining pits. As in other villages, the company planted cashew nuts in the rehabilitated areas, whereas the claims for the collection of cashews were unclear. Owing to their inability to speak French, the population has not formulated its claims versus CBG in a coherent way to date (Interview with Elders 3/02/2014).

Boundouwouande

Boundouwouande is a particularly severe case of mistreatment of the local population. The village is encircled by mining sites and is only accessible by a steep narrow dirt track. Any car that wants to enter the village has to be escorted by the mining companies. Together with harsh living conditions, this led to recurrent blockings of the surrounding mines by the population. Since 2005, the local authorities have thereby pushed for a resettlement, but CBG instead tried to sedate the villagers by building a school, two wells and a health post:

"They gave three class rooms, they are no pupils. [...] A nice school with three class rooms over there is not used. They gave a health post to this small village: there is no doctor. They gave a well drilling for a population of less than 30 people. All of that was a waste of resources, a useless waste of resources that we have denounced in every respect. But as this is a company that does not belong to us, we have denounced, but it is CBG alone which is responsible for that. [...] If they have to give something to the community, they just have to conduct studies, [...] ask the population: what are your problems? What do you need? If somebody explains to you what he needs, if you give him what he wants, you help him. But you can't give a ship in Sangaredi, there is no sea!" (Interview with Administrator Sangaredi 28/01/2014; translation by JK).

Neither the school nor the health post is functional, because the building activities were not coordinated with the local administration and did not respect the PDL (Interview with Consultant 30/01/2014; Administrator Sangaredi 28/01/2014; Administrator Sangaredi 3/02/2014). One of the two wells has not been working for 3 years (Interview with Elders 2/02/2014; Consultant 30/01/2014). The size of the village is highly contested: while the locals stated that the village comprises 125–150 families and count 500 inhabitants altogether (Interview with Elders 2/02/2014), the administration in Sangaredi maintained that no more than 50–60 people are living in Boundouwouande (Interview with Administrator Sangaredi 3/02/2014; Administrator Sangaredi 3/02/2014). Similar disagreements happened during my visit of Ndanta Fonye and were resolved in favor of the local population. In addition, in this case, the administration believed that the village only counted about 100 inhabitants, but the villagers convinced him that this only concerned the family heads. Subsequently, he revised his estimations to 1000–1500 inhabitants.

When I entered the village of Boundouwouande, I was shocked by the amount of dust in the air. All villagers had red eyes. Their complaints were similar to those of other villages around the mines, but their needs seemed to be much more urgent as the dust rendered agriculture unviable and even the domestic animals did not survive due to the blastings and a lack of nutrition. Their most important demand, namely, the construction of a road—primarily to sell cashew nuts and coal in town—had not been implemented. Only one villager was employed by CBG, as a trash collector (Interview with Elders 2/02/2014). The terms put forward by the villagers in case of a resettlement were, nonetheless, quite extensive: they demanded houses in town with water and electricity, a school, a mosque, and jobs for those who have a profession (Interview with Elders 2/02/2014).

When I asked the *chef de service du développement rural* of Sangaredi, who was in charge for this failure at CBG, he mentioned the Service environnement and the *Relations communautaires* (Interview with Administrator Sangaredi 28/01/2014). Both sections are situated in Kamsar. I first tried to talk to executives in Sangaredi and was directed to the head of the section *Santé, sécurité et environnement* who also declared himself responsible for *Relations communautaires*. For the sake of intelligibility, I will call this executive after the Guinean football player Idrissa Sylla. Sylla answered my questions in an anxiously snappy manner, but claimed to have been circumvented in the process in Boundouwouande. He heavily criticized the building projects and maintained that they were implemented by executives from Kamsar with helping hands in Sangaredi, but could not or did not want to give any names (Interview with Administrator Sangaredi 28/01/2014). A few days later in Kamsar, I met the superintendent of the section *Projets communautaires* of CBG, whom I will call Ismaël Bangoura, another Guinean football player. When I asked about Boundouwouande, Mr. Bangoura denied that Idrissa Sylla was in charge of such projects; instead, he only had a "comptroller for the evaluation of the sites"[143] in Sangaredi (Interview with Executive CBG 18/02/2014). When I asked again if he knew Mr. Idrissa Sylla, Bangoura began to talk to a student trainee who had been allowed to listen to the interview as if I were no longer in the room:

> "BANGOURA: He talks about [Sylla] and I don't know him. I know him as [Sylla], but I don't know him in the project, no, not at all. If he constructs schools, all the better!
>
> KNIERZINGER: Yes, there was a small problem, monsieur, because…
>
> BANGOURA: There is no problem! There is no problem! […] This project does not exist here! Do you understand? It does not exist! […] Fine. That's why I say that one has to avoid mixing things up. When I say avoid mixing things up, what do I mean by that? That's when, what I don't know, you believe in what people say. Anything that is not written down: No. This project, I don't know it. [Sylla], I don't know him, ok?
>
> KNIERZINGER: I got it, monsieur, I got it.
>
> BANGOURA: Fine" (Interview with Executive CBG 18/02/2014; translation by JK).

A few days later, I was lucky enough to meet a superior of Ismaël Bangoura who behaved in a less authoritarian manner and who finally clarified the situation. According to the directeur santé *sécurité environnement et incendie* of CBG, CBG had even commissioned an impact study on Boundouwouande in 2002/2003 that was implemented by the Guinean company BERKA, which has its office in Sangaredi. When I asked him if he had a copy, he stated that it should be in Sangaredi. When I then told him that Idrissa Sylla in Sangaredi also did not have any copies, he called in Sangaredi, but did not succeed in localizing the study. He also thought that the responsible executive was Ismaël Bangoura, or his superiors of the *Direction administrative*. When I told him about my assignation with Bangoura, he called him in front of me and confronted him with his perspective. Bangoura finally told him that the construction work (a school, a health post, and two wells) had been made by the Service engineering that had no authority to do so. The *directeur santé sécurité environnement et incendie* concluded the following:

"[Ismaël Bangoura] confirms that he had not been associated and that the "engineering" has done that. [...] They have denounced it because the engineering should in principle work under the orientation of the "communities". But the engineering has taken the project and has constructed without associating them. But they have protested afterwards. [...] As this was a project more or less within the framework of the... the new... the new plateau that will be exploited, thus, when the impact study was finished, there was indeed a plan that it would be good to have a school in this village. Thus, the engineering has taken its part, they made the plan and they executed it without asking the community. That's why the people have protested that it's not normal to usurp their power. Voilà, that's the situation. But that's really isolated, that's an isolated case. In all the other community projects, they have the budget. The "community", [Bangoura], he has a large budget per year. Thus, they take care of that" (Interview with Executive CBG 23/02/2014; translation by JK).

The Construction Site of GAC

Global Alumina Corporation started to work in an area west of CBG's claims in 2000 (Interview with Executive GAC 1/02/2014). In 2012, the company employed 83 direct and about 700 indirect employees (Chambre des Mines de Guinée 2012: 46). GAC is a typical junior company that entered the country with particularly high ambitions concerning social and ecological standards, but was eventually bought out by more potent investors. After an intermezzo with BHP Billiton—which completely quit Guinea in 2012—the US company was taken over by investors from the United Emirates, Mubadala, and Dubai Aluminium. Rather than the promised refinery in Sangaredi, these investors are currently about to construct a refinery in the Emirates that will buy considerable amounts of CBG's bauxite and will initially use all of the bauxite produced by GAC. By the time of writing, the scheduled start of mining in Sangaredi was 2018. However, the Ebola outbreak is likely to further postpone this date. The start of the refinery has no deadline at present and will probably not be realized at all (Interview with Sociologist GAC 1/02/2014; Executive GAC 1/02/2014).

Construction work in Sangaredi has not yet started, with the exception of two new villages for relocated farmersof the area and a harbor in Kamsar. The harbor had been partly financed by the United States African Development Foundation (Diallo 2005b: 10). This institution has been "established to support African-led development that grows community enterprises by providing seed capital and technical support [and] provides grants of up to $250,000 directly to hard-to-reach and underserved community enterprises that are ready to do their part" (USADF 2015). GAC was at that time US-owned and Diallo stated that it received 1 million USD for the construction of the harbor (Diallo 2005b: 10).

The relocation of about 500 people (45 families) in 2008 in the area, where the refinery was to be constructed (Interview with Executive GAC 1/02/2014)[144], was accompanied by impressive social and environmental efforts. Following detailed impact studies and consultations of the population (cf. Knight Piésold and Co. 2008), the company constructed 20 well drillings, six schools, four health posts, three "multi-purpose centers" (*Centres polyvalents*), a car bridge, and a mosque (Interview with Executive GAC 1/02/2014). Trees and plantations that existed in the two relocated villages have been partly reproduced in the new settlements and the villagers were both compensated for their loss of income and are free to use the old plantations as long as they exist. Only the houses of the villages were destroyed. GAC's compensations for plantations are about ten times higher than those of CBG (Interview with Administrator Sangaredi 28/01/2014). In addition to these payments for lost property and income, the villages also received the so-called *compensations communautaires*: money that was to be distributed to all inhabitants and not only to owners of land and property (Interview with Executive GAC 1/02/2014). The amount of compensation paid to women was not disclosed to their husbands and fathers to increase their financial independence (Manson and Knight 2007). Probably, the most important measures comprised capacity building programs and the formation of social groups, covering the amelioration of agriculture and livestock production, alphabetization, professional trainings, such as car and lorry driving, and reforestation). All of these activities had the explicit aim to create capacities that would not be lost in case of a failure of the project, as well as stimulating local economic dynamics aside from mining (Interview with Executive GAC 1/02/2014). The community superintended of GAC conceded that this task is quite difficult to attain, given that the population primarily wants employment: "It's easier to have a monthly salary in the pocket than to work on a long-term perspective with much smaller revenues" (Interview with Executive GAC 1/02/2014; translation by JK). This is why GAC also tried to stimulate local agriculture by building up local providers of vegetables for the factory itself:

"It's our vision to try to develop exactly this possibility of connecting the activities that are around the site and the supply: why isn't there a poultry farm around, where one could take them directly? And everything also depends on the philosophy of the investor. If you have an investor who is afraid to become ill because he does not trust the sanitary service in the country, [...] he cannot allow himself to nourish all his workers who could become ill afterwards, and so on. In the moment, we cannot rely on the state to assure us, for instance, that the meat is fit for consumption. There is no service like that. The state also has to do something" (Interview with Executive GAC 1/02/2014; translation by JK).

The reforestation measures of GAC have also set new standards. The company founded groups that produce a variety of local seeds that can be used to at least partially recreate the biosystem of former mining pits. CBG is also invited to buy these seeds (Interview with Administrator Sangaredi 28/01/2014). In 2014, these various activities were also extended to the town of Sangaredi. This happened due to the interventions of the local authorities (Interview with Administrator Sangaredi 30/01/2014) and based upon the argument that many workers of GAC actually lived in town. The late interest for the town is also related to GAC's main "competitor":

"Above all, we always thought that CBG was over there and that's why it's a bit... hm. But there is, nonetheless, great need down there which is still not met and that's why this year we have decided to invest quite a lot in this area" (Interview with Executive GAC 1/02/2014; translation by JK).

This statement once again shows the huge gap between the official public procedures that involve the elaboration of development plans and the rationalities of the companies, who ultimately decide when to build which installations. In 2014, GAC was about to construct a new school with six class rooms and rehabilitated an old school with 24 classrooms, a lycée and a health post (Interview with Executive GAC 1/02/2014). The *Délégué scolaire de l'enseignement élémentaire* of Sangaredi additionally mentioned the construction of an incinerator, an operating room and a renovation of the laboratory of Sangaredi's health center and the installation of a dozen well drillings in schools and districts in town (Interview with Administrator Sangaredi 29/01/2014). Altogether, GAC's budget for community spending amounts to 2.5 million USD, reflecting about 50 times the local budget fully controlled by the Guinean state in this area (cf. calculations on Kamsar in the former section). Since 2004, the company has invested 25 million USD for ecological and social purposes in Guinea (Interview with Executive GAC 1/02/2014).

In terms of communication with the local political authorities, the *chef de service du développement rural* of Sangaredi conceded that GAC was "far ahead of the others" (Interview with Administrator Sangaredi 28/01/2014). Both local politicians and village authorities have been frequently consulted and involved into the companies' considerations. The general view of GAC on politics is, nonetheless, rather pragmatic and shortsighted. While the company has received a 15-year tax holiday (Manson and Knight 2007), it is neither engaged in capacity building programs for the local administration nor does it involve governmental representatives in financial matters. The company respects all provisions and even supports the authorities to do this, for instance, in the case of the local development plans (PDL). However, financial control and the award of contracts remain within the company (Interview with Executive GAC 1/02/2014). When I asked a senior executive of GAC if this was due to a lack of local capacity or rather corruption, she remarked:

"I think there are people who have the capacity to manage, but we have no guarantee that the funds would be used properly. I have been working in the field of international development for more than 20 years and when we started the situation was catastrophic, the funds that passed through… the governments. If you give that to the government, what will remain for the population? They don't receive anything" (Interview with Executive GAC 1/02/2014; translation by JK).

From the view point of this executive and most likely also GAC, the company seemed to face a trade-off between efficiency (or material well-being) and the democratic control of resources (or self-determination). As efficiency seems to be higher estimated than democratic control, the company does not follow the example of several gold mining companies or the tentative steps of CBG to implicate local authorities in fiscal matters. In view of the current change of investors, this not only constitutes one of the few problems of the local administration with GAC (Interview with Administrator Sangaredi 30/01/2014), but calls the sustainability of all the companies' efforts into question. One way to increase democratic control as much as possible was to create *Comités de suivi environnemental et de la réinstallation*, which comprise local political authorities and administrators from Sangaredi and Boké. This case was particularly paradoxal, as the company seemed to have insisted upon its own control:

"These people just refuse to… to come. We have to go to them to tell them what we have done. And they look and they ask questions. They never come to us! It's their role to come to the site to […] He doesn't do his work. Why doesn't he do it? Lack of resources. They have no money. We have to pick them up. We have to pick them up with our own cars. We have to give them money so that they come. Thus, we say: Ok, we come over to you instead! [*imitates the supervisors*] But for us to come to the reunion, you have to give us something! We cannot work like this! It's your role to make sure that what GAC does in this country… You have to supervise! […] It's your role, you are paid by the state to do this. It's your country, […] it's your population! You cannot do as you please! You should supervise us! But you want us to give you money to supervise us? That won't work" (Interview with Executive GAC 1/02/2014; translation by JK).

This lack of control is also evident in Fria and Kindia. In Fria, the mayor complained that Rusal did not reforest as provided, but conceded that this was also a problem of governmental control (Interview with Mayor of Fria 18/01/2014). In Kindia, I met the directeur régional des mines who is also responsible for the control of the reforestation measures of Rusal in the region. He was unable to show me the legal provisions for reforestation or point to a single person outside the company realm who actually controlled these measures (Interview with Administrator Kindia 17/03/2014). According to a former employee of Friguia, Rusal even had paid environmental taxes, at more than 15,000 USD per year for environmental protection and 1 million USD per year for producing chemical substances. Apparently, none of these taxes have found their way back to Fria (Interview with Engineer 23/01/2014).

Similar concerns also emerged regarding the employment policy of GAC. At first, the company tried to fulfill its promise to employ mainly people from the mining areas by simply recruiting workers at the site. Given that this often led to nepotistic practices, the company created local committees. However, these committees rather worsened the situation as they led to bribery and were not accepted by the local authorities. Now the company has lists of potential candidates from the concerned villages and selects them randomly.

With the retreat of the US investors in 2012, the population around the site of GAC in Sangaredi risks suffering the same fate as their colleagues in many other mining towns: without public control over the infrastructure, most installations and institutions will cease to be operational within a few years. A senior executive pointed out that the alumina project is not to start before 2018—if ever—and the demand for bauxite is not about to rise in the years to come. These meager prospects are worsened by the current Ebola outbreak. A senior executive of the *Lab biomedical* of CBG already reported in 2014 that the relocated population was not content with their new homes and felt "more or less abandoned" (Interview with Executive CBG 3/02/2014). As the alumina factory will most likely not be erected, their homes have been destroyed for nothing. This means that—despite its considerable ambitions—GAC also failed in substantially increasing Guinea's share of value added and did not succeed in diversifying its economy. The "historic turning point for the country"—as a GAC representative put it in 2007—is thus yet to come (Committee on foreign affairs 2007: 45; see also Diawara 2014: 63).

3.3 Summary of the Chapter

3.3.1 Post-Colonial Exploitation: Infrastructural Power Replaces Despotic Power

The production networks of Rusal and CBG are two examples of transnational decision spaces that are entangled with the formal political space while at the same time undermining it. These conflicting forms and strategies of control neither lead to new consolidated spaces nor do they create new consistent forms of order. Due to their transnational nature and the variety of equally powerful actors in play, power configurations in the mining towns change rapidly and strongly depend on restructurations of the global aluminum industry. On the one hand, international mergers or emergency sales can render hundreds of thousands of people dependent on food aid, as the case of Fria has shown. On the other hand, the financial stability of these TNCs can lead to "islands of prosperity" (Soumah 2008: 215) within one of the poorest countries in the world, as the case of Sangaredi shows to a certain degree. However, also in the latter case, the spatial and temporal limits of these islands of prosperity are evident and—in contrast to their limited success in terms of material well-being—self-determination and nationwide economic development rather suffered from these semi-corporate constellations of rule. Former territorial regulation under the French colonizers—which was already very weak compared with European nation states—has neither been replaced by an effective Guinean state apparatus nor by non-territorial forms of transnational (corporate) governance, but rather by quite arbitrary forms of rule, which are characterized by their high grade of non-discursive or direct control.

On the one hand, the examples of Rusal and CBG have shown that both the conflicting sets of Guinean law (different mining codes, mining conventions, other Guinean law bodies, such as the labor code and customary law), and international regulations, such as companies' codes of conduct or the WTO, provide ample possibilities for rescaling, jeux d'échelles, and brokerage. Until the mining code of 2011, mining conventions were negotiated on an ad hoc basis, providing different fiscal incentives to each firm. For instance, in the case of CBG, the state receives 65% of the profits while only possessing 49% of CBG's stakes (International Monetary Fund 2008: 9).

On the other hand, the high infrastructural power of the companies in these cities often renders these negotiations about norms almost senseless and superficial. Based upon the control of railroads, electricity, pipe water, health care, etc., company representatives exert direct control, which does not need any other justification to be exercised. For instance, in the case of Fria, the conflictual situation showed that Rusal did not need to refer to any legal or value-based framework anymore. By withdrawing its Russian specialists, stopping the influx of petrol that fired the city's power stations inside the factory and without even clarifying the status of the Guinean workers, the company both managed to economize its spending during the current economic downturn and still exercised sufficient control to pressure the government into giving out further stakes in Guinea (Dian-Dian), again without fulfilling promises to restart the factory to date. A similar—almost physical—dependence exists in Sangaredi and Kamsar and can also be followed down the chain of command outside the official corporate realm, for instance, to the mostly illiterate local population around the mines. The situation in the highly capital-intensive bauxite towns of Guinea, therefore, seems to be strongly driven by coercive power: while infrastructural power also depends on social acceptance, the circle of technicians and managers who can virtually switch these Guinean cities on or off is quite small.

The elements of infrastructural power in the four analyzed mining towns have already been listed several times in this book. A synopsis of these two company realms shows many similarities between Fria, Sangaredi, and Kamsar. Débélé is too small and too far away from the regional capital Kindia to have a decisive influence on the life of the city dwellers (Interview with Préfet of Fria 19/01/2014), but also in this case, thousands of families are likely to face hardship in the next years to come when the bauxite mines will be depleted, because neither the Guinean government nor Rusal have socio-economic plans for the settlement after the retreat of Rusal. As in the other three Guinean bauxite towns, almost all the infrastructure projects, all of the maintenance and most social services are provided by the private company. This was also made explicit by the government, which until recently refused to extend local development measures and capacity building programs (mainly under the *Programme d'appui aux communautés villageoises*) to the mining regions by pointing to the investments of the companies (cf. Diallo et al. 2011: 52). The use of public buildings thereby mostly depends on decisions of the mining directions rather than the local authorities. The same applies for electrification, tap water, housing, firefighting, sewer cleaning, waste disposal, medication, and even for the maintenance of the passenger train between Sangaredi and Kamsar (e.g., only CBG has the cranes that can carry away derailed cars). Calculations on infrastructural projects in Kamsar showed that less than one % of public infrastructure development is actually carried out and controlled by the local authorities, whereas national public expenditure (which is highly dependent on taxes from the aluminum companies) almost exclusively comprises salaries (Interview with Activist 2/03/2014). The few nationally coordinated infrastructural projects—mostly financed and thereby controlled again by external donors (cf. USAID 2006: 3; Dux 2000)—are typically carried out in non-mining regions. Local authorities thereby become mere spectators—or in most cases, critical commentators—of the central political processes in the mining towns. The population is more or less at the mercy of global aluminum prices: cost reductions in times of less demand naturally result at first in cuts of voluntary corporate "responsibility" measures, as can be seen in the history of all mining towns (cf. Alfa Issa Thiam in Soumah 2008: 143).

Moreover, the local media are thoroughly controlled by the companies. In particular, the local radio stations—Radio CBG for Sangaredi and Kamsar and Rusal's *Voix de Fria*—reach a large percentage of the population. While Radio CBG seems to be quite strictly apolitical (Interview with Journalist 13/04/2014), Rusal has founded its radio only in summer 2011, at a time when the tensions with the trade unions could no longer be talked away (Deguenon and Traoré 2011c: 8). The prefect of Fria called the radio a "cement for the reinforcement and the consolidation of the social fabric" and encouraged the city administration to "continue in this regard and to work for social peace and socio-economic and cultural development in our city" (ibid; translation by JK). However, a series of about 15 h of emissions on the conflict in Fria (Interview with Journalist 19/01/2014), called "Zoom sur l'usine" rather underlined the perspective of Rusal during the crisis. The "coordinator" of the radio was an expatriate (Deguenon and Traoré 2011c: 8).

Security, schooling, entertainment, and religious facilities are only partly maintained by corporate investments. CBG has more or less successfully outsourced several hotels, bars, schools, and small businesses during structural adjustment in the 1980s and 1990s. In addition to private security forces, the companies directly employ several policemen and soldiers in the mines and exchange foodstuff, housing, petrol, and other gifts for the loyalty of a large part of the public security forces (Interview with Police Commissioner 17/02/2014). The case of Fria showed that schooling, entertainment, and religious life also considerably suffered from the shutdown the factory, mainly due to missing financial inflows through workers' salaries. The most compelling indicator for an overall decline of living standards in Fria is, without doubt, the considerable decline in rice consumption per head since Rusal's departure. A small survey among street vendors showed that rice consumption for this group declined by about one-third after the stop of the factory. In addition, former workers and other interviewees stated that they only eat once rather than twice a day since the start of the crisis. Similar scenarios are probable for the other mining towns once the mines will be closed. This particularly concerns Débélé, whose mine will be closed within the next 5 years. Given that the Guinean government has lacked the required know-how to construct and maintain the existing infrastructure in the mining towns, the corporatization of public services probably had many positive consequences. Without this new constellation of rule, the few people who had been living in these areas before the arrival of aluminum majors would continue with small-scale farming or fishing or migrate to other urban accumulations. They would probably have no access to schools, wards, electricity, or tap water. However, my case studies showed that most of the inhabitants of the bauxite company towns still do not benefit from these developments and there also seems to be some evidence that especially the fragmentation of decision space has had negative long-term consequences. Similar to the paradox of the plenty on the national scale, the separation of formal political representation and actual political practice leads to a dependence on company structures also in terms of social organization. Shortly after Rusal had left Fria, people began to dispose and burn their wastes next to their houses and even next to the hospital. The only initiative to tackle this problem came from the former vice director of Fria, who was still living in the city. In one of his recurrent appearances in the local radio, he even gave the impression that this engagement was part of his responsibility as the former vice director (Interview with Journalist 19/01/2014). A village near Kamsar that was concerned by waste deposal of CBG even refused to talk to the sub-prefect and demanded to speak directly to the decision makers, who were thought to sit in the so-called "White House", the head office of CBG in Kamsar (Interview with Executive ANAIM 17/02/2014). The most important impulses to reinstall social services in Fria came from a small group of young educated political and economic entrepreneurs, mostly sons and daughters of former workers, who formed the NGO ARSYF to attract private aid, in many cases from other *Friakas* abroad. However, such as many other actors, this interest group tended to confuse corporate governance with government. One of its recent projects was the foundation of a radio in Fria, which several of its members described as "private: belonging to the civil society" (propre à la société civile; Interview with Activists 15/03/2014). The Guinean activist Moktar Diallo describes the situation as follows:

"You have the impression that the administration becomes an appendix of the private enterprise. [It] transforms into an agent of sensitization and mediation when there are conflicts between the private sector and the population. What does all this to the citizen? [...] The state is senseless, is not existing, is an illusion. The real deciders, the real powers are the guys who have the factories, the guys who run the mines. Thus, it's them who have to be addressed. [...] Sometimes you see the prefect much more often in the factory than at his office. You have therefore the increasing absorption of the public administration by the private administration" (Interview with Activist 20/01/2014).

Diallo thereby described a vicious circle of perceived and "real" political influence: the lower the perception of political influence, the less local political authorities are inclined and allowed to interfere, which in turn again reduces their popularity and their ascribed capability to change things (cf. Larrue 1997: 271). The companies' explications for their dedemocratized form of "public" spending remain quite paternalistic and also show the primacy of their own codes of conduct over national law and the aspirations of the political class in Guinea. All interviewed company representatives responded that money would only be wasted if they would not retain control. In the case of CBG, this not only encompassed voluntary CSR but even local taxes. Until the end of the 1990s, CBG paid its local taxes to the prefectural authorities in Boké, who were subsequently responsible for the distribution to sub-prefectural authorities (Diallo et al. 2011: 52). As this did not work out, the company simply began to directly carry out programs in the sub-prefectures. The few local taxes in Kamsar and Sangaredi are thereby paid out "in kind" in the form of infrastructure realizations and social programs.[145] Similar arguments can also be discerned regarding the rule of law. Rather than remitting a car accident that killed a villager to the courts, the prefect of Fria tried to find an "amicable solution" between the company and the family, which resulted in simple bargaining about the price of the dead (Interview with Administrator Fria 5/04/2014). This bargaining approach was also predominant in the other cases of the destruction of property or the infliction of bodily harm.

Besides negative experiences concerning the financial control of sub-national authorities (which were without doubt also sanctioned by higher tiers of the company structure outside Guinea), the reluctance of the aluminum majors to build up and trust in local capacities is also related to concrete interest constellations: the company officers in charge of infrastructural projects seem to live quite well off their responsibilities. In addition, Western aluminum companies, such as Pechiney, have a quite long history of state critical ideology that received a new push with the advent of CSR (cf. Chambre des Mines de Guinée 2011: 16, 31). Even in *comparatively* well-functioning democracies of OECD countries, these companies increasingly portray themselves as ecological and cost-effective alternatives to allegedly outdated state institutions.

In addition to these political effects, many development measures of the companies are not coherent. In the case of CBG, various sections were involved in community spending when I conducted my field research. On several occasions, the officers of one section were not aware of the projects of the other sections, or at least pretended not to be. Many interviewees argued that these non-transparent and nested structures also breed substantial corruption. In Fria, the frequent changes of factory ownership since the 1990s particularly led to the frustration of rural communities affected by mining; as successive impact studies carried out by the French, American and Russian companies never led to actual projects.

3.3.2 Status Quo in Guinea

Despite practically all global players of the aluminum sector competing for concessions since the turn of the millennium, the Guinean GDP per capita has fallen and poverty has risen in the last 10 years. This situation has been further aggravated by the global economic crisis—reducing both the price for aluminum and the long-term prospects of the involved companies—and the Ebola outbreak (Diallo et al. 2011: 7, 19, 26). At the time of writing, Guinea's real food prices are among the highest in the world. Oxfam only lists Zimbabwe and Angola as countries with less affordable foodstuffs (Oxfam 2015). Since the lockout in Fria, less than 5000 people are employed in the bauxite sector, subcontractors included. The Guinean power supply remains very poorly developed, despite the huge potentials of the country and renewed announcements during the 2000s (see Touré 2009: 96f). Contrary to the renewed great expectations, even processing in Fria (which refined about 10–15% of Guinean bauxite) has stopped, leaving the country with only a tiny fraction of the value added of the TNCs (cf. Haefliger 2012).

The window of opportunity of the 2000s is about to close. One important indication is the retreat of BHP Billiton from Guinea in early 2012 when the company sold both its stakes in Guinean iron ore projects and its share of GAC in Sangaredi (Rusal 2012). This drawback is certainly not only related to the global economic situation, but also to the corruption scandals, the political uncertainty, and large appetite of the Guinean government for fast money. Besides "saving" money for the time after political rule, the subsequent governments also had to find money to stay in power. As Guinea did not qualify for the HIPC Initiative until fall 2012[146] and the new mining projects did not get ahead, the government had to spend most of its budget for the refunding of debts. The World Bank did not allow Guinea to take new loans before being able to repay the old debts (Interview with Consultant 10/02/2012). In order to fulfill at least a few of their promises made before the elections, the rulers resorted to solutions, such as the Rio Tinto deal in April 2011 to fill the Guinean treasury. The company agreed to pay a lump sum of 700 million USD only to obtain the mining concession at Mount Simandou. Until today, the legal basis for this transaction remains unclear (Permatasari 2012; Consultant 10/02/2012).

With Guinea's qualification for the HIPC Initiative and the cancellation of parts of Guinea's debts, this situation only gradually changes. The government budget will continue to come mainly from four enterprises (Alcoa, Rio Tinto, Rusal, and AngloGold Ashanti) and from foreign grants and funds. This means that sovereign political decisions remain quite rare in Guinea. The political economist Bonnie Campbell directly links the "capacity of Guinea to introduce its own development strategies [...] to access to revenues from the bauxite sector", which in turn depends on (1) "terms and prices on which mineral resources are sold on world markets" (Campbell 2009b), (2) the capacity for local transformation, and (3) linkage effects to other sectors. All these points have been part of development plans since the 1960s, but in no case, progress was made (ibid.). Today, the price for bauxite is again on the decline, political stability is not (yet) achieved (see the announcement of a putsch just before the last elections and Dadis Camara's running at the presidential elections in October 2015; cf. Châtelot 2015), the consequences of the global financial crisis still limit the willingness of industrial countries to attain their goals of development cooperation, and Ebola still scares off international investors.

Notes

1. A census in 2014, 2 years after the closing of the factory in Fria, counted 61,697 inhabitants (République de Guinée 2014).
2. These are figures of the last census in 2014. Other sources came to much higher figures (see Sect. 3.2).
3. Campbell (2009b), Journalist 23/02/2012, MBendi Information Services (2013a), World Investment News (2009). For Aredor, see Diallo (2009), Fofana (2011), MBendi Information Services (2013b).
4. Brüggmann (2007), Investguinée (2012), Khrennikov (2012), Kennedy (2008), Revoin (2012).
5. BBC News (2012), Diallo (2010, 2013l, 2014l), Former executive Rusal 25/03/2014, Emejo (2014), Metal Bulletin (2013), Odey (2013), Rusal (2014b), Aminata (2013).
6. See Bah (2012a), Husband et al. (2009: 56f), International Monetary Fund (2012a: 67), Soumah (2008: 231f).
7. Diallo (2012m, 2014f), Khrennikov (2012), Neue Zürcher Zeitung (2010).
8. Beside the municipality, this prefecture comprises three sub-prefectures, namely, Baguinet, Tormelin, and Banguiny. The mine also streches out to Baguinet and Tormelin (Interview with Activist 2/03/2014).
9. Before the lock out in 2012, the city was much larger. In January 2014, the mayor of Fria estimated that the whole prefecture counted 125,000 inhabitants (Interview with Mayor of Fria 18/01/2014). The results of the census in 2014—published in April 2014—registered only a population of 97,527 (République de Guinée 2014: 7). This difference can already be related to the lock out of Rusal.
10. Some interviewees also stated that several religious buildings still received electricity (Interview with Journalist 17/01/2014).
11. The shipping from Fria to Cameroon and Spain was undertaken by the firm SANAGA, belonging by half to Aluminium Pechiney (Debost 1977: 4; Pechiney Ugine Kuhlmann 1977: 4).
12. Since the mid-1990s (Balzli and Herbermann 2000; OEC 2015).
13. Former executive Rusal 4/04/2014, Former executive Rusal 25/03/2014, Holloway (1988), see also Campbell (1991: 34–35), Lanning and Mueller (1979: 370).
14. Lake Volta, the largest artificial lake by surface, covers 8500 km^2.
15. This comprised the current territories of Republic of the Congo, Gabon and the Central African Republic.
16. This text specifically concerned the employees of Alucam.
17. Marc-Edmond Morgaut, Secrétaire Général du Comité Supérieur de la Promotion Sociale en Algérie, distinguished between promotion sociale for the élites and for the masses.

18. The first of these facilities was in Salin-de-Giraud, a formerly deserted area that necessitated the construction of a small town for the mostly foreign workers. All workers were provided with housing, medical care, and education programs (e.g., knitting and cooking workshops for the workers' daughters) and were also encouraged to attend church (see Fommervault and Vindt 2001).
19. On corporatism, see Schmitter and Lehmbruch (1979).
20. Today, this would amount to about 3.3 billion USD in exports and to taxes of about 250 million USD in these 13 years of operation. This is rather low compared to the taxes paid by CBG in the following years.
21. MARG was dissolved December 1959 (Larrue 1997: 376).
22. In fact, it did that via another institution, namely, the Comité d'aménagement régional de Guinée. MARG was set up mainly as a research unit based on Conakry that should map the area and conduct sociological and ecological studies—in the beginning also for Boké—although the funding for this site ceased immediately after independence. A third body—the Fonds d'aménagement régional de Guinée—was to create the social, administrative, and physical infrastructure of Fria and attenuate the negative effects of industrialization (Pauthier 2002: 56) .
23. Pauthier did not find an exact breakdown of these debts (Pauthier 2002: 87).
24. Cf. Soumah (2008: 84), Pauthier (2002: 69), Larrue (1991: 61).
25. The French Fonds d'Investissements pour le Développement Économique et Social financed mainly infrastructural and agricultural projects and supported industrial enterprises.
26. These recruitment tests began already in 1956 under the leadership of Marc-Edmond Morgaut (Larrue 1997: 81–84; Pezet et al. 2009: 17).
27. Larrue also quotes Jean-Paul Alata—a member of the French communist party—who had been working as a financial advisor of Touré before being detained in the Camp Boiro and who stated in 1977 that "besides Fria and CBG, everything is just a fiasco" (Larrue 1997: 143; translation by JK).
28. Before the retreat, the insurgents also burned down Touré's villa Belle Vue, a present of Western Germany in its recognition struggle against the GDR, in which they expected him to have hidden (Marx 1977: 4).
29. In addition, there was also the National Congress, "meeting at least once in 5 years" to elect the Central Committee, theoretically the central ruling body of the party. However, the committee was usually overpowered by its executive, the National Political Bureau (Adamolekun 1976: 25).
30. They comprise the "conspiracy of April 1960", the so-called "teachers' plot" in November 1961, the "traders' plot" in 1965, and the "plot of

the military" in 1969. The party organized trials against the people who were said to have plotted against the regime, although the accused persons were never present and the judgments were always death sentences, inprisonment for life, or hard labor (Kaba 1977: 30; Johnson 1970: 357; Schicho 2001: 328ff).

31. Cf. Trade unionist 17/04/2014, Larrue (1997: 181, 203), Trade unionist 28/03/2014, Former executive Rusal 7/04/2014.
32. According to Bah, employees of Fria received about 5–10,000 syli a month. Equally qualified civil servants earned 5–6000. A sack of rice could be purchased for 1000 syli by the workers of Fria. Others had to pay 2–3000 (syli Politician 28/03/2014). According to Laparra, these cheaper provisions, mainly of cereal, oil, and sugar, more than doubled the worker's purchasing power in the 1950s and 1960s (Larrue 1997: 107). Several retirees in Fria who had been blue-collar workers stated that they had earned only 3000 syli in the 1980s.
33. To this end, the partners formed the "Alumina Company of Guinea" (Diallo et al. 2011: 35).
34. Deguenon and Traoré (2010: 1), Diallo (2013f, k, m, 2014f), Storozhenko (2009).
35. Currency conversions have been made with the calculator of OANDA and are based upon the exchange rates of April 2014, unless otherwhise stated (see OANDA 2015). Calculations of inflation are based upon World Bank data in the case of Guinea (World Bank 2015) and on the website CoinNews for rough calculations in USD (CoinNews 2015).
36. Balde (2012), Diallo (2010: 6), 2012l, o, 2014f, Guinéenews (2015), Storozhenko (2009), Tilouine (2014).
37. Diallo (2005b: 3, c: 5, d: 6), Guinéenews (2015). When a delegation of these investors wanted to visit the facilities, they were prevented from entering by the personnel of Rusal (Diallo 2005a: 4).
38. See also Diallo (2012c, 2014g).
39. A Guinean dentist in the company hospital replaced an expatriate when he began working in 1987 after having been trained in Algeria. According to him, his salary was one-tenth of his predecessor's (Politician 28/03/2014).
40. Some of the translators were Russians and were fully employed. The Guinean translators were not (Diallo 2014g).
41. Diallo even recounts that the director Anatoly Pantchenko escaped a lynching in a workshop during and because of this restructuration (Diallo 2012k).
42. Former executive Rusal 4/04/2014; Entrepreneur 7/04/2014; Storozhenko (2009).

43. Interview with Entrepreneur 7/04/2014; Former executive Rusal 4/04/2014; Activist 3/04/2014; Worker Rusal 3/04/2014.
44. Deguenon and Traoré (2011f), Diallo (2012c), Kadyrov (2011), Rusal (2011c).
45. I was unable to obtain exact figures on Russian expatriates' wages. The salaries of Guinean civil servants are regrouped into three categories. The ceiling for the lower category is at 1.5–1.7 million GNF (about 200–250 USD), while the average earns 700–800,000 GNF (100–120 USD). The salary of ministers is called prime de fonction (Interview with Trade unionist 17/04/2014).
46. The picture shows a river that never dried out and had fish before the start of the blastings. The pillar in the middle was used to cross the river during the rainy seasons.
47. Former executive Rusal 4/04/2014; Chef de District Koumi-Kimbely 7/04/2014; Former executive Rusal 7/04/2014; Administrator Fria 1/04/2014; Administrator Fria 5/04/2014.
48. Journalist 15/01/2014; cf. Deguenon and Traoré (2011b: 5), Huy (2013).
49. Trade unionist 19/01/2014, Brygo (2009), Diallo (2013k, 2014g); Mayor of Fria 18/01/2014.
50. Given that all company documents are still locked in the factory, it was difficult to verify Diallo's figures. The annual publications of the Guinean mining chamber do not support his claim, although this could be because Russians were employed as subcontractors but received the same or even better fringe benefits (cf. Chambre des Mines de Guinée 2011, 2012). On this basis, even the actual number of core workers (employees) became obscure.
51. Diallo (2012d, e, i, n, 2013e, 2014f).
52. Commission des 132 agents de Rusal Friguia (2012b), Diallo (2012i, 2014q); cf. Diallo (2013c, d, e, 2014f).
53. Qualified workers of CBG (they have finished so-called Ecoles professionelles to become electricians, mechanics, etc.) received a basic salary of 2.4 million GNF or the equivalent of 200 kg bread per month.
54. Burgis et al. (2010), Diallo (2010: 30, 2012d), Neue Zürcher Zeitung (2010).
55. Sidibé (2013a, f, g, h, i).
56. The worker's tribunal declared the strike illegal shortly thereafter, on 18 April 2012 (Diallo 2013e).
57. Diallo (2012n, 2013e, 2014f), Interview with Activist 14/03/2014, Sidibé (2013b).
58. The census of 2014 counted 61,697 inhabitants for *Fria-Centre* and 96,527 inhabitants for the whole prefecture of Fria. However, these

numbers were considerably higher before the stop of the factory. According to the 2012 annual report of the public hospital in Fria, *Fria-Centre* counted 89,348 inhabitants (Interview with Hospital director 1/04/2014). Amara Traoré, the mayor of Fria, Ibrahima Diallo and another informant therefore rather tend to speak of 120,000 concerned inhabitants (Diallo 2014t; Journalist 15/01/2014; Journalist 17/01/2014; Mayor of Fria 18/01/2014). In an article of the company newspaper in 2010, the director of Friguia estimated the urban population "at more than 70,000" (Kadyrov 2010: 3).

59. According to Barry, 80% of the married workers are polygamous (Barry 2008: 20). In 1997, Larrue quoted a study indicating that the average worker's family in Fria comprised two wifes and eight children (Larrue 1997: 245).
60. Diallo (2013b, e, 2014g, r), Mayor of Fria 18/01/2014.
61. In September 2012, Rusal announced that it would allow for the evacuation of another 13 cases of emergency until the end of the year (Kadyrov 2011: 1–2).
62. Cf. Interview with Activists 15/03/2014; Activist 3/04/2014, Diallo (2013b, 2014q), Journalist 19/01/2014.
63. Contrary to the perception of most of the interviewees, the available figures on mortality in Fria do not show any augmentation with the stop of the factory. The public hospital registered 40 deaths in 2013, compared to 39 in 2011 (Interview with Hospital director 1/04/2014). The death registry of the municipality even shows a decline of mortality from 29 cases in 2011 and 2012 to 2015 cases in 2013. This registry is considered relatively trustworthy, because it is connected to the payments of rents to the relatives (Interview with Administrator Fria 7/04/2014). If these numbers are correct, the only explanation is migration to other cities of more than half of the workers since April 2012.
64. Unfortunately, I was not allowed to conduct interviews in the company hospital.
65. Association des Ressortissants et Sympathisants de Fria (2012: 10), Interview with Activists 15/03/2014, Diallo (2013l), Administrator Fria B 30/03/2014, Mayor of Fria 18/01/2014.
66. Cf. Administrator Fria A 30/03/2014; Mechanic 4/04/2014; Administrator Fria B 30/03/2014; Engineer 18/01/2014.
67. Interview with Former executive Rusal 4/04/2014; Former executive Rusal 7/04/2014; cf. Activist 3/04/2014.
68. They officially ended the strike on 1 June 2012 (Diallo 2012i, 2014q).
69. Diallo (2012h, 2014f, q), Trade unionist 19/01/2014; Diallo (2012h, 2014k), Trade union leader 3/03/2014, Sidibé (2013a).

70. Diallo (2012b, m, 2014h, q); Sidibé (2012b), Mayor of Fria 18/01/2014.
71. Diallo (2012h, 2014k); Trade union leader 3/03/2014; Sidibé (2013a).
72. Trade unionist 17/04/2014; Former executive Rusal 3/04/2014, Trade union leader 3/03/2014, cf. Diallo (2012a, 2013b, 2014f), Journalist 15/01/2014.
73. Barry (2013a), cf. Diallo (2013b, e, g, 2014i, r).
74. I asked first how much rice was consumed per month and then how many people ate from this rice.
75. Cf. Trade unionist 18/01/2014; Engineer Rusal 2/04/2014; Street vendor 1/04/2014; Engineer Rusal 4/04/2014.
76. Already in August 2012, Rusal sent 2.5 tons of rice to the "conseil des sages" of Fria, along with medicaments and small allowances for the guards at the factory. In addition, the government also sent rice, apparently also paid by Rusal (Diallo 2013b, 2014f, q). When Lansana Kouyaté—the former prime minister of the transition government before the coup d'état of Dadis Camara—wanted to enter the city in August 2012 with food, hospital beds and drinking water, the government blocked the streets (Diallo 2013b, 2014q; Sidibé 2012a). Kouyaté had been pushed into the government of Lansana Conté by the social movement of 2007 under the leadership of the trade unions. In April 2014, the NGO ARSYF distributed about 5 tons of rice donated by US members of the organization (Interviews NGO Reunion 29/03/2014). This had happened at least four times since the start of the crisis in April 2012 (cf. Diallo 2014s). Two months later, the government sent another 20 tons of rice to Fria (Diallo 2014g), probably in connection with the World Food Programme of the United Nations, which started to provide Fria with food in July 2014 (Diallo 2014o; see also Diallo 2013l; Femmes de Fria 2013; Journalist 19/01/2014).
77. Sidibé (2012c), Diallo (2012b, 2014f, q), Journalist 19/01/2014, Journalist 17/01/2014.
78. Bah and Diallo mention a group of 20 people who were still employed to guarantee a minium service (Interview with Activists 15/03/2014). The missing hundred could be security forces.
79. "[I]l n'y a pas eu vol, mais du pillage" (Diallo 2014g).
80. Diallo (2014g, r), Sidibé (2013d, e), Journalist 19/01/2014.
81. Diallo (2013i, 2014g, r), Sidibé (2013c).
82. "Nous ne pouvons rien contre Rusal, propriétaire de l'usine, il est plus puissant que l'Etat." Cf. Diallo (2014q), Journalist 19/01/2014, Sylla (2013).
83. Several interviewees called this contract the "Accords de Paris" (Interview with Police 28/03/2014; Diallo 2014n).

84. Diallo and Kaba (2014), Tilouine and Té-Léssia (2014), Diallo (2014e, j).
85. Interview with Former executive Rusal 4/04/2014; Préfet of Fria 19/01/2014; Activist 3/04/2014; Activist 2/03/2014; Activist 14/03/2014; Former executive Rusal 25/03/2014; Employee Fria 6/04/2014.
86. "Une radio qui est privée: propre à la société civile." This is interesting both with regard to the positive connotation of the term "private" and concerning the—privately owned—radio that already exists. While the members of ARSYF see the state as a powerless institution that is not capable to "protect" them (Interviews NGO Reunion 29/03/2014), Rusal here again figures as a non-private entity.
87. Interview with Trade unionist 19/01/2014; Former executive Rusal 25/03/2014; Journalist 3/04/2014; Engineer Rusal 4/04/2014; Activist 3/04/2014.
88. This is the last available figure for the commune urbaine of *Kindia*-Centre. The whole prefecture of Kindia counts 438,000 inhabitants (République de Guinée 2014: 11).
89. In 2013, Diallo talked about 2 billion USD (Diallo 2013c).
90. Helmer (2005); on Nikolaiev see also Trade union leader 3/03/2014, Delasnerie and Diallo (2004), Journalist 23/02/2012, Legal expert 13/03/2014.
91. Dux (2000), Trade unionist 25/03/2014, Soumah (2008: 152–153), République de Guinée 2013c: 7, Employee CBK 18/03/2014.
92. Campbell (1986: 31–33, 1991: 35, 2009b: 77–78), Dux (2000), Habig (1983: 87), Holloway (1988: 45).
93. Campbell and Clapp (1995: 435), Dux (2000), Trade unionist 25/03/2014, République de Guinée (2013c: 7).
94. Only in 2013, the Guinean supreme court ruled that they had been wrongly convicted (Interview with Trade union leader 3/03/2014).
95. Interview with Administrators Mambia 18/03/2014, Employee CBK 18/03/2014, Diallo et al. (2011: 52), Executive CBK 20/03/2014, Former Sous-Préfet of Mambia 24/03/2014; Activist 18/03/2014.
96. Administrators Mambia 18/03/2014, Sous-Préfet of Mambia 18/03/2014, Compagnie des Bauxites de Kindia 2013: 4,7–11, Rusal (2014d: 10–11, 2010b: 4), Administrator Mambia 19/03/2014, Rusal (2011b: 11), Administrator Kindia 17/03/2014.
97. The bauxite train cannot be used by the population.
98. Interview with Activist 17/03/2014; Executive CBK 20/03/2014; Haefliger 2012; Administrators Mambia 18/03/2014.
99. When Rusal took over, its chief executive promised a basic wage of 500 USD that should have be attained in 2011. However, it still amounts

to no more than 200 USD at present (Interview with Executive CBK 21/03/2014).

100. Employee CBK 18/03/2014; Commission des 132 agents de Rusal Friguia 14/12/2012b; Trade unionist 25/03/2014.
101. In October 2014, Rio Tinto annouced the sale of its aluminum smelter in Cameroon (Investir au Cameroun 2014).
102. This is the outcome of the last census in 2014 (République de Guinée 2014: 7), although several interviewees mentioned much higher figures. According to the sub-prefect of Kamsar, his administrative division counts 365,800 inhabitants (Interview with Sous-préfet of Kamsar 12/02/2014). The DSE of Kamsar 1 and other informants even talked about 500,000 inhabitants (cf. Soumah 2008: 141; Administrator Kamsar 10/02/2014). However, an expert of the Guinean National Institute of Statistics who had taken part in the census confirmed this comparatively low figure.
103. Only 6 years after the mineral bauxite received its name from the French village Les Beaux by the geologist Pierre Berthier in 1821, the French explorer René Caillié found the precious material near the settlement of Boké in Guinea, from where he started his famous expedition to Timbuktu (Soumah 2008: 123).
104. Dux mentions initial reserves of about 10 million tonnes of bauxite on the Loose Islands (Dux 2000). Already in the late-1970s, Guinean annual production amounted to more than 10 million tons (cf. Soumah 2008: XII).
105. According to CBG, Bauxites du Midi had already been a French affiliate of Alcan in 1920 (Johnson-Sahr 1991: 5; République de Guinée 2013c).
106. Campbell (1986: 26–29, 1991: 34, 2009b: 77), Larrue (1997: 132), N'Diaye 1984: 25, République de Guinée (2013c), Vuerings (1964: 98).
107. See Der Spiegel (1971).
108. Soumah (2008: 123), cf. Compagnie des Bauxites de Guinée 1983: 10, République de Guinée 2013c: 12.
109. The relations between the US and Guinea improved accordingly when Kennedy came into power in January 1961. During the Cuban Missile Crisis, Touré even refused to open up the Conakry airport as an operational base for Soviet military and received assistance from USAID, the US Exim Bank and the Overseas Private Investment Corporation (Leimgruber 1990: 219–232; Soumah 2008: 143–150; République de Guinée 2013c: 13; Soumah 2008: 147).
110. Holloway (1988: 33–36), Soumah (2008: 124–125), cf. Johnson-Sahr (1991: 8), République de Guinée (2013c: 13).

111. Employee CBG 14/02/2012; Compagnie des Bauxites de Guinée 2012. This was confirmed by Vladimir Mamedov, a Russian geologist in Sangaredi (Interview with Executive 13/02/2012).
112. These about 6000 boats constitute about 80% of all bauxite that had been produced by CBG until 2011.
113. This proportion fell again to 22.95% in 2007, when the Canadian Alcan was taken over by the British-Australian Rio Tinto.
114. Initially, the price of bauxite had been fixed at 0.5% of the price of aluminum ingot. However, owing to the high stakes of CBG, the government then demanded one percent instead of the stipulated 0.5% from CBG. In the end, CBG paid 0.75% (Soumah 2008: 171).
115. Today, the purchasing power of 200 million USD in 1977 would amount to about 750 million USD in the US (cf. US Department of Labor 2014). Overall governmental income from mining was less than 370 million USD in 2012 (cf. EITI 2015; for more details see Campbell 1991; République de Guinée 2013c: 27; Soumah 2008: 181).
116. Soumah (2008: 171–174), cf. Compagnie des Bauxites de Guinée (1983: 14), Forster (1976: 135).
117. Diallo (2005d: 5), Commission des 132 agents de Rusal Friguia (2012a), Trade unionist B 25/02/2014.
118. This article has been reformulated during the revision process and became less clear in the last version of the mining code.
119. The Guinean Code du Travail also regulates subcontracting (Bah 2014a).
120. Translation by JK.
121. A similar situation prevails in Conakry. According to the World Bank, only half of the 13,000 "consumers" pay for the electricity that they use. In 2006, the World Bank financed a project to identify the housholds which receive electricity (Samb 2006: 18).
122. Sociologist 22/02/2014; Executive CBG 18/02/2014; Sous-préfet of Kamsar 12/02/2014.
123. Sous-préfet of Kamsar 12/02/2014; Soumah 2008: 141; Mayor of Kamsar 17/02/2014.
124. Administrator Kamsar 13/04/2014; Sociologist 22/02/2014; Administrator Kamsar 10/02/2014.
125. Cf. Barry (2013b: 10), Executive CBG 18/02/2014, Administrator Kamsar B 13/02/2014, Consultant 21/02/2014.
126. Translation by JK.
127. One of the most interesting measures in recent years has been the establishment of micro-enterprises or TPEs (Très petites entreprises). Since 2010, CBG has created at least seven such enterprises with various specializations, such as construction and maintenance of buildings,

sanitation and cleaning, reforestation, cleaning of industrial gutters, and the production of jam (République de Guinée 2013c: 25). These measures could hold great importance in terms of reducing the impact of a possible stop of production as it happened in Fria, as well as creating greater value added in the region. However, as the mentioned examples show, this can also be used by the company as a cost reduction instrument, whereby work previously been undertaken by employees can be delegated to poorly paid and uninsured "self-employed" or contract workers. For instance, the current construction of new workers' houses in Sangaredi uses day laborers who are paid below the minimum wage.

128. The cooperation with the responsible person at CBG, the superintendent of the section *Projets communautaires*, was particularly difficult. He did not allow photographs to be taken of documents and "answered" many questions by mechanically reading out company regulations without allowing interposed questions.
129. E.g., Interview with Activist 2/03/2014; Executive CBG 18/02/2014; Administrator Kamsar B 13/02/2014; Consultant 21/02/2014.
130. The competent executive of CBG maintained that the company always considered the PDL for these independent projects. However, given that these projects do not enter the sub-prefectural budget until now, this statement is rather superficial (Interview with Executive CBG 18/02/2014).
131. GAC also maintains a harbor in Conakry and will have to transport its bauxite by train. The neighboring communities of these facilities will thereby also profit from a small part of this budget. However, this percentage spent on other communities is surely equalized by the community spending of Rusal and Henan Chine from which I was not able to get any figures.
132. The prefectural budget in Boké almost exclusively consists of wages (Interview with Consultant 21/02/2014) and investments from the national level did not exist in the latest years apart from the PACV.
133. CBG 2012; Doctor Kassappo 19/02/2014; Sous-préfet of Kamsar 12/02/2014.
134. Barry (2013b: 24), Doctor Kassappo 19/02/2014, Christian leader 14/02/2014.
135. Executive CBG 25/02/2014, Executive CBG 23/02/2014, République de Guinée (2013c), Legal expert 13/03/2014.
136. The census of 2014 counted 76,538 inhabitants for the sub-prefecture of Sangaredi (République de Guinée 2014: 7. Figures of CBG were slightly higher (cf. Employee CBG 14/02/2012; Compagnie des Bauxites de Guinée 2012) and administrators at the sub-prefectoral level counted more than 150,000 inhabitants for the whole

sub-prefecture (Interview with Consultant 30/01/2014; Leno 2014a: 2; Administrator Sangaredi 3/02/2014).

137. Consultant 30/01/2014; Administrator Sangaredi 28/01/2014; Administrator Sangaredi 3/02/2014.
138. According to the Guinean ministry of health, the construction of a health post costs 600–700 million GNF, compared to about 1.4 billion GNF for a centre de santé (health center) and about 5 billlion GNF for a *centre de santé amelioré* (Interview with Administrator Conakry 26/03/2014).
139. Cf. Interview with Administrator Sangaredi 29/01/2014; Activist 29/01/2014.
140. Probably together with the taxes paid by Henan Chine. Touncara stated that he did not have contact with the Chinese company. Their taxes are collected by the directeur régional des mines in Boké, who transfers them to Sangaredi (Interview with Administrator Sangaredi 3/02/2014).
141. Executive CBG 10/02/2014; cf. Administrator Mambia 20/03/2014; Executive CBG 23/02/2014; Engineer CBG 6/02/2014.
142. Minerals always belong to the state, regardless of whether the land in question is privately owned or not.
143. Translation by JK.
144. Contrary to this source, Knight Piésold & Co. report on the resettlement of a "total of 164 households and 782 individuals" (Knight Piésold and Co. 2008). This number probably also contains resettlements for the construction of the harbor.
145. Administrator Kamsar B 13/02/2014; Administrator Sangaredi 30/01/2014; cf. Diallo et al. (2011: 52).
146. See Barry (2006b: 9), Diallo et al. (2011: 85), Diallo (2012b, 2014q), International Monetary Fund (2012b).

References

Adamolekun, Ladipo. 1976. *Sekou Toure's Guinea*. London: Methuen.

Agnew, John, and Richard Grant. 1996. Representing Africa: The Geography of Africa in World Trade, 1960–1992. *Annals of the Association of American Geographers* 86 (4): 729–744.

AidData. 2015. Exploration of aluminum in Boke district, Guinea. http://china.aiddata.org/projects/228, checked on 6/08/2014.

Alcoa. 2015. Corporate Overview. http://www.alcoa.com/global/en/about_alcoa/overview.asp, checked on 21/03/2015.

Aminata. 2013. Rusal a réduit les pertes au 3ème trimestre après la fermeture des usines non rentables. In aminata.com, 13/11/2013. http://aminata.com/

rusal-a-reduit-les-pertes-au-3eme-trimestre-apres-la-fermeture-des-usines-non-rentables/.

Arieff, Alexis, and Nicolas Cook. 2009. *Guinea's 2008 Military Coup and Relations with the United States*. Washington, DC: Congressional Research Service.

Arieff, Alexis, and Nicolas Cook. 2010. *Guinea: Background and Relations with the United States*. Washington, DC: Congressional Research Service.

Arieff, Alexis, and Michael McGovern. 2013. "History is stubborn": Talk About Truth, Justice, and National Reconciliation in the Republic of Guinea. *Comparative Studies in Society and History* 55: 198–225.

Association des Ressortissants et Sympathisants de Fria. 2012. Rapport des résultats d'enquête sur la scolarité des enfants des travailleurs de Rusal-Friguia. Projet de mobilisation des organisations de la societé civile pour le plaidoyer autour de la crise de l'usine Rusal-Friguia. Fait à Conakry le 19 septembre 2012. Fria.

Bah, Boubacar Sidy. 2006. *Etude des déperditions de soude de la société RUSAL/FRIGUIA et leurs impacts sur l'environnement. Mémoire du diplôme de fin d'études supérieures*. Conakry: Université Gamal Abdel Nasser.

Bah, Mamadou Oury. 2012a. Mines: Rusal et l'Etat guinéen s'accordent sur l'exploitation de Dian-Dian! In *joursdeguinee.com*, 31/12/2012. http://joursdeguinee.com/fichiers/blog12.php.

Bah, Khalirou. 2013. Ibrahima Soumah, ancien ministre des mines et de l'enseignement technique à Guinée. In *Guinéenews*, 10/04/2013. http://www.guineenews.org/detail_article.asp?num=20134107282, checked on 15/05/2013.

Bah, Abdoulaye. 2014a. Hémicycle: le Conseil national de la transition adopte un nouveau code de travail attrayant. In *guineenews.org*, 11/01/2014. http://guineenews.org/hemicycle-le-conseil-national-de-la-transition-adopte-un-nouveau-code-de-travail-attrayant/.

Balzli, Beat, and Jan Dirk, Herbermann. 2000. Berüchtigte Kundschaft. In *spiegel.de*, 29/05/2000. http://www.spiegel.de/spiegel/print/d-16525418.html, checked on 16/10/2014.

Balde, Mamadou Sinna. 2005. *L'Histoire des relations entre la Guinée et le fonds monétaire international de 1963 à 2002. Memoire de Maitrise*. Conakry: Université Gamal Abdel Nasser.

Balde, Ciré. 2012. La CNTG exige la résiliation du contrat de Rusal et de VALE|Vision Guinee. In *visionguinee.info*, 27/12/2012. http://www.visionguinee.info/2012/12/27/la-cntg-exige-la-resiliation-du-contrat-de-rusal-et-de-vale/, checked on 6/08/2014.

Banque mondiale. 2007. Guinée—Ressources naturelles en Guinée: les mines de l'espoir? Washington.

Barry, Alpha Ousmane. 2004. *Les racines du mal guinéen*. Paris: Karthala.

Barry, Lamarana Gocky. 2008. *Etude psychologique d'une entreprise: Cas de l'usine de Fria. Memoire*. Kankan: Université Julius Nyerere.

Barry, Aminatou. 2006a. Village Communities Support Program (PACV) for Participative Decentralized Development: Results Beyond Expectations. *La Revue—Magazine semestriel du bureau de la banque mondiale en Guinée* 1: 31–34.

Barry, Tabara Sakho. 2006b. After So Much Sacrifices, Our Populations Deserve More Solidarity From the International Community. *In La Revue—Magazine semestriel du bureau de la banque mondiale en Guinée* 1: 6–11.

Barry, Mamadou. 2012. CBG ou "l'incompétence et la médiocrité au sommet de la gestion du personnel". In *Guinee libre*, 23/07/2012. http://www.guineelibre.com/article-cbg-ou-l-incompetence-et-la-mediorite-au-sommet-de-la-gestion-du-personnel-108440344.html, checked on 17/05/2013.

Barry, Diao. 2013a. Mines: Russal rafle l'annexe 11 de Dian-Dian. In *lejourguinee.com*, 6/03/2013. http://www.lejourguinee.com/index.php/fr/economie/973-mines-russal-rafle-l-annexe-11-de-dian-dian, checked on 6/08/2014.

Barry, Dramé Aicha. 2013b. La CBG abrite les travaux du Comité Technique. *CBG-News* 80 (2013): 5.

BBC News. 2012. Nigeria court strips Russia's Rusal of Alscon aluminium plant. In *bbc.com*, 10/07/2012. http://www.bbc.com/news/world-africa-18781717, checked on 13/08/2014.

Bierschenk, Thomas, and Jean Pierre de, Olivier Sardan. 2002. Local Development Brokers in Africa. The Rise of a New Social Category. Department of Anthropology and African Studies Working Papers Vol. 13. Mainz: Johannes Gutenberg University.

Bornefeld, Andreas. 2007. Russland—Oligarchen: Oleg Deripaska—Rusal. In *netstudien.de*, 10/12/2007. http://www.netstudien.de/Russland/deripaska.htm, checked on 12/08/2014.

Boullet, Daniel. 2000. Pechiney and the Environment, 1960–1980. In Les Cahiers d'histoire de l'aluminium 26.

Bouliguine, Alexandre. 2007. "Il ne faut pas se reposer sur les lauriers". *In La voix de Rusal* 25 (4): 1–2.

Brinkmann, Manfred. 2004. Silber aus Lehm. *Wie nachhaltig ist Aluminium?* Aachen: Brot für die Welt.

Brüggmann, Mathias. 2007. Oligarch Oleg Deripaska: Wie ein Bulldozer. In *handelsblatt.com*, 27/04/2007. http://www.handelsblatt.com/unternehmen/management/koepfe/oligarch-oleg-deripaska-wie-ein-bulldozer/2801134.html, checked on 13/08/2014.

Brygo, Julien. 2009. Guinea's bauxite town falls apart. Who benefits from Russian-African coproduction? In Le Monde diplomatique, 11/2009.

Bureau de Presse de la Présidence. 2013. Bauxite: La Guinée veut tripler sa capacité de production d'ici à 2020, 17/03/2013. http://aujourdhui-en-guinee.com/fichiers/blog15-999.php?idc=fr_Bauxite__La_Guin_e_veut_tripler_sa_capacit__de_production_d_i&PHPSESSID=05e44b9f8290f93ac0cd947a746c48aa, checked on 2/07/2014.

Burgis, Tom. 2010. Behind the wrangle for Guinea's minerals. In *ft.com*, 5/07/2010. http://www.ft.com/intl/cms/s/0/2e4ec1be-8853-11df-aade-00144feabdc0.html, checked on 23/10/2014.

Burgis, Tom, Mitchell, Tom, and Catherine, Belton. 2010. Guinea demands share of UC Rusal offering. In *ft.com*, 26/01/2010. http://www.ft.com/intl/cms/s/0/d2bb3c8a-0a1a-11df-8b23-00144feabdc0.html, checked on 23/10/2014.

Camara, Lanciné. 2006. ACG/Friguia privatisé. In Horoya 6402, 5/05/2006.

Camara, Ismaël Kabiné. 2008. Guinée: émeute sanglante à Mambia. In *Afrik.com*, 11/10/2008. http://www.afrik.com/article15412.html, checked on 28/09/2014.

Camara, Al Hadj Ibrahima. 2012. Un acte significatif du syndicat. In La voix de Rusal 58, p. 9.

Camara, David. 2013. Allocution du Directeur Général Adjoint de Rusal, Monsieur CAMARA David, suite à la mission conjointe avec une délégation des employés auprès de la représentation de RUSAL. In *guineemining.info*, 23/04/2013. http://www.guineemining.info/index.php?id=285&tx_ttnews%5Btt_news%5D=7417&tx_ttnews%5BbackPid%5D=283&cHash=f1a6fb976c, checked on 12/08/2014.

Camara, Alia. 2014. Kamsar: des affrontements entre civils et forces de sécurité font 1 mort et plusieurs blessés. http://www.factuguinee.com/fichiers/videos5.php?idc=fr_Kamsar__des_affrontements_entre_civils_et_forces_de_s_curit__, checked on 16/06/2014.

Campbell, Bonnie K. 1983. Les enjeux de la bauxite. La Guinée face aux multinationales de l'aluminium. Montréal: Les Presses de l'université de Montréal.

Campbell, Bonnie K. 1986. *Bauxite Bargaining. The Consequences for a Bauxite-Producing Country of the Renegotiation of the Bauxite Levy: The example of Guinea*. Vienna: United Nations Industrial Development Organization.

Campbell, Bonnie K. 1991. Negotiating the Bauxite. Aluminium Sector Under Narrowing Constraints. *Review of African Political Economy* 51: 27–49.

Campbell, Bonnie, and Jennifer Clapp. 1995. Guinea's Economic Performance Under Structural Adjustment: Importance of Mining and Agriculture. *The Journal of Modern African Studies* 33 (3): 425–449.

Campbell, Bonnie (ed.). 2009a. *Mining in Africa. Regulation & Development*. Ottawa: International Development Research Centre.

Campbell, Bonnie K. 2009b. Guinea and Bauxite-Aluminium: The Challenges of Development and Poverty. In *Mining in Africa. Regulation & Development*, ed. Bonnie Campbell, 66–119. Ottawa: International Development Research Centre.

CBG. 2012. Santé: la CBG au chevet du Centre de Santé de Kassapo. http://www.cbg-guinee.com/actions-communautaires/155-sante-la-cbg-au-chevet-du-centre-de-sante-de-kassapo, updated on 15/06/2014, checked on 16/06/2014.

CECI. 2014. Guinée: Renforcement des communautés rurales. http://www.ceci.ca/fr/regions/afrique/guinee/projets-fr-CA/renforcement-des-communautes-rurales-projet-termine/, checked on 17/11/2014.

Chambre des Mines de Guinée. 2011. La Revue de la Chambre des Mines de Guinée. Conakry.

Chambre des Mines de Guinée. 2012. La Revue de la Chambre des Mines de Guinée. Conakry.

Cherie FM Guinée. 2013. Grève illégale des employés: La Cour d'Appel de Conakry a tranché en faveur de Rusal/Friguia. In *cheriefmguinee.com*, 30/04/2013. http://www.cheriefmguinee.com/fichiers/livre12-999.php?langue=fr&code=calb6052, checked on 12/08/2014.

CoinNews. 2015. Inflation Calculator. http://www.usinflationcalculator.com/, checked on 3/02/2015.

Commission des 132 agents de Rusal Friguia. 2012a. De la prise en charge des agents arbitrairement mis à la retraite de Rusal Friguia. A ce jour virtuels retraités. Letter to Directeur general de la caisse nationale de securité sociale. Fria, 8/12/2012. Data file.

Commission des 132 agents de Rusal Friguia. 2012b. de notre rejet de la Société pour motif de soi-disant retraites. Letter to Premier Ministre Mohamed Saïd Fofana. Fria, 14/12/2012. Data file.

Committee on foreign affairs. 2007. *Prospects for Peace in Guinea. Hearing Before the Subcommittee on Africa and Global Health of the Committee on Foreign Affairs.* Serial No. 110–32. Washington, DC: US Government Printing Office.

Compagnie des Bauxites de Guinée. 1983. *Working Together in Guinea.* Pittsburgh: CBG; OFAB.

Compagnie des Bauxites de Guinée. 2012. Welcome at Sangaredi. A reference bauxite mine since 1973. Powerpoint presentation of CBG for visitors. Sangaredi.

Compagnie des Bauxites de Kindia. 2012. Rapport de l'augmentation du rendement de travail et du salaire moyen.

Compagnie des Bauxites de Kindia. 2013. Pour le développement durable. Powerpoint slides.

Conakry Infos. 2014. Poutine contraint le PDG de Rusal, Oleg Deripaska à ouvrir une usine de ciment. In *youtube.com*. https://www.youtube.com/watch?v=Q2szcdHGIn4, checked on 18/04/2014.

Condé, Alpha. 1972. *Guinée: Albanie d'Afrique ou néo-colonie américaine?.* Paris: Editions Git le Coeur.

Custers, Raf, and Ken, Matthysen. 2009. *Africa's Natural Resources in a Global Context*. Antwerp.

Dadco. 2013. Dadco Alumina & Chemicals Ltd. http://www.dadcoalumina.com/, checked on 17/05/2013.

Debost, Jean. 1977. La guenon, la locomotive et la bauxite. *Aluminium Pechiney* 31: 4–9.

Deguenon, Edgar, and Ibrahima Sory, Traoré. 2010. Friguia: Un demi-siècle d'histoire. *In La voix de Rusal* 40 (2): 1.

Deguenon, Edgar, and Ibrahima Sory, Traoré. 2011a. Crise Financière: Comment se porte le marché de l'aluminium aujourd'hui? *In La voix de Rusal* 57 (12): 4.

Deguenon, Edgar, and Ibrahima Sory, Traoré. 2011b. Déclaration officielle. *In La voix de Rusal* 57 (12): 5.

Deguenon, Edgar, and Ibrahima Sory, Traoré. 2011c. LA VOIX DE FRIA: Vers une nouvelle dynamique? *In La voix de Rusal* 57 (12): 8.

Deguenon, Edgar, and Ibrahima Sory, Traoré. 2011d. L'aluminium a plongé sous la barre des 2000 Dollars. *In La voix de Rusal* 57 (12): 2.

Deguenon, Edgar, and Ibrahima Sory, Traoré. 2011e. Pour le respect strict des horaires de travail. *In La voix de Rusal* 56 (11): 6.

Deguenon, Edgar, and Ibrahima Sory, Traoré. 2011f. RUSAL/FRIGUIA: Direction et syndicat au chevet de notre hôpital. *In La voix de Rusal* 48 (3): 6–9.

Deguenon, Edgar, and Ibrahima Sory, Traoré. 2011g. RUSAL/FRIGUIA: Le salamac des pèlerins à la Direction générale. *In La voix de Rusal* 57 (12): 11.

Deguenon, Edgar, and Ibrahima Sory, Traoré. 2011h. RUSAL/FRIGUIA: Un comité de médiateurs pour désamorcer la crise. *In La voix de Rusal* 57 (12): 6.

Delasnerie, Sébastien, and Akoumba, Diallo. 2004. Guinea: Transforming a geological scandal into an economic miracle? In *Aluminum International Today*, 11/2004. http://img.docstoccdn.com/thumb/orig/139876710.png, checked on 16/05/2013.

Der Spiegel. 1971. Gift im Bier. In *spiegel.de*, 1/02/1971 (6), p. 75. http://www.spiegel.de/spiegel/print/d-43732553.html, checked on 10/01/2012.

Der Standard. 2014. Oligarch Deripaska stockt bei Strabag wieder auf. In *derstandard.at*, 12/06/2014. http://derstandard.at/2000001948570/Oligarch-Deripaska-stockt-bei-Strabag-wieder-auf, checked on 12/06/2014.

Deripaska, Oleg. 2014. Quand la discipline s'impose à la métallurgie. *In La voix de Rusal* 76: 3.

Deutsche Welle. 2013. Bauxitabbau in Débélé. Deutsche Welle, 5/2013.

Diallo, Akoumba. 2005a. L'Aurore. In L'Aurore 55, 25/10/2005.

Diallo, Akoumba. 2005b. L'Aurore. In L'Aurore 56, 8/11/2005.

Diallo, Akoumba. 2005c. L'Aurore. In L'Aurore 57, 28/11/2005.

Diallo, Akoumba. 2005d. L'Aurore. In L'Aurore 58, 13/12/2005.

Diallo, Akoumba. 2009. Aredor, annulation de la concession Minière, 7/04/2009. http://www.laurore.info/index.php?id=400&tx_ttnews[pointer]=26&tx_ttnews[tt_news]=6196&tx_ttnews[backPid]=160&cHash=d47041a546, checked on 11/10/2013.

Diallo, Aboubacar Akoumba. 2010. Redevances minières à Fria. Fria: Projet de renforcement des capacités des organisations de la société civile et de la jeunesse guinéenne.

Diallo, Mamadou Chérif, et al. 2011. *Les enjeux de la gouvernance du secteur minier en Guinée*. Conakry: Cooperation Guinée Allemagne.

Diallo, Aboubacar Akoumba. 2012a. Complexe des bauxites et alumine de Dian Dian. In Le Lynx 1053, 18/06/2012.

Diallo, Amadou Kendessa. 2012b. La crise de l'usine Rusal-Friguia: Le cauchemar des Friakas est loin de connaître son épilogue. In *lejourguinee.com*, 3/10/2012. http://www.lejourguinee.com/index.php/fr/nouvelles/politique/7-la-crise-de-l-usine-rusal-friguia-le-cauchemar-des-friakas-est-loin-de-connaitre-son-epilogue, checked on 8/08/2014.

Diallo, Ibrahima Talibé. 2012c. Retour vers le néant pour Fria usine—ville, Rusal à Fria. In *guineemining.info*, 27/03/2012. http://www.guineemining.info/index.php?id=285&tx_ttnews%5Bpointer%5D=1&tx_ttnews%5Btt_news%5D=3076&tx_ttnews%5BbackPid%5D=283&cHash=400f378c43, checked on 4/08/2014, checked on 5/08/2014.

Diallo, Ibrahima Talibé. 2012d. Cher Lecteur. In *guineemining.info*, 5/04/2012. http://shop.radio-kankan.com/index.php?id=285&tx_ttnews%5BpS%5D=1328050800&tx_ttnews%5BpL%5D=2505599&tx_ttnews%5Barc%5D=1&tx_ttnews%5Bpointer%5D=84&tx_ttnews%5Btt_news%5D=3157&tx_ttnews%5BbackPid%5D=283&cHash=66ab7591c3, checked on 6/08/2014.

Diallo, Ibrahima Talibé. 2012e. Fria le 03 Avril 2012. In *guineemining.info*, 5/04/2012. http://shop.radio-kankan.com/index.php?id=285&tx_ttnews%5BpS%5D=1328050800&tx_ttnews%5BpL%5D=2505599&tx_ttnews%5Barc%5D=1&tx_ttnews%5Bpointer%5D=84&tx_ttnews%5Btt_news%5D=3157&tx_ttnews%5BbackPid%5D=283&cHash=66ab7591c3, checked on 5/08/2014.

Diallo, Ibrahima Talibé. 2012f. Fria le 27 avril 2012: Encore une victoire de plus pour Rusal. In *guineemining.info*, 29/04/2012. http://www.guineemining.info/index.php?id=285&tx_ttnews%5Btt_news%5D=3389&tx_ttnews%5BbackPid%5D=283&cHash=7bec50cf03, checked on 5/08/2014.

Diallo, Ibrahima Talibé. 2012g. La probable nième victoire de Rusal, où la fin de la glorieuse ville de Fria. In *guineemining.info*, 10/05/2012. http://www.guineemining.info/index.php?id=285&tx_ttnews%5Btt_news%5D=3450&tx_ttnews%5BbackPid%5D=283&cHash=e7f12da4a1, checked on 6/08/2014.

Diallo, Ibrahima Talibé. 2012h. Amères situations à Fria la situation de Fria est très bizarrement. In *africavision.info*, 16/05/2012. http://www.africavision.info/index.php?id=350&tx_ttnews%5Bpointer%5D=85&tx_ttnews%5Btt_news%5D=3508&tx_ttnews%5BbackPid%5D=349&cHash=d854a0ec53, checked on 6/08/2014.

Diallo, Ibrahima Talibé. 2012i. Rusal victorieuse en Guinée. In *guineemining.info*, 5/06/2012. http://www.guineemining.info/index.php?id=285&tx_ttnews%5Btt_news%5D=3677&tx_ttnews%5BbackPid%5D=283&cHash=cea645315f, checked on 6/08/2014.

Diallo, Ibrahima Talibé. 2012j. Rusal contribue a la dépravation socio-économique de Fria, et détruit la guinée. In *guineemining.info*, 21/11/2012. http://sanakou.press-guinee.com/index.php?id=285&tx_ttnews%5Btt_news%5D=5407&tx_ttnews%5BbackPid%5D=282&cHash=85274fb94f, checked on 8/08/2014.

Diallo, Ibrahima Talibé. 2012k. Quel avenir pour Fria après l'alumine et l'hécatombe de Rusal? In *guineemining.info*, 27/11/2012. http://www.guineemining.info/index.php?id=285&tx_ttnews%5Btt_news%5D=5456&tx_ttnews%5BbackPid%5D=282&cHash=cd088f7a31, checked on 8/08/2014.

Diallo, Ibrahima Talibé. 2012l. Mines: Rusal a réussi ses vœux. In *guineemining.info*, 24/12/2012. http://www.guineemining.info/index.php?id=285&tx_ttnews%5Btt_news%5D=5811&tx_ttnews%5BbackPid%5D=282&cHash=a2635ea35d, checked on 6/08/2014.

Diallo, Ibrahima Talibé. 2012m. Fria: Rusal doit quitter la Guinée. In *guineemining.info*, 12/12/12. http://sanakou.press-guinee.com/index.php?id=285&tx_ttnews%5Btt_news%5D=5593&tx_ttnews%5BbackPid%5D=283&cHash=5fe576419a, checked on 5/08/2014.

Diallo, Ibrahima Talibé. 2012n. Fria le 13 avril 2012 Rusal au sommet de sa gloire si cela en était une. In *africavision.info*, 14/04/2012. http://www.africavision.info/index.php?id=350&tx_ttnews%5Bpointer%5D=96&tx_ttnews%5Btt_news%5D=3241&tx_ttnews%5BbackPid%5D=349&cHash=c02217ea21, checked on 5/08/2014.

Diallo, El Hadj Mohamed. 2013a. Mines: L'AMC compte dès 2016, transporter sa première tonne de bauxite. In *guinee7.com*, 25/06/2013. http://www.guinee7.com/component/content/article/37-communiques/3149-mines-lamc-compte-des-2016-transporter-sa-premiere-tonne-de-bauxite, checked on 6/09/2013.

Diallo, Ibrahima Lelouma. 2013b. Fria: La naïveté des populations en jeux pour être politiquement au service des contradictions nationales et ne pas être un enjeu pour la 3éme république et les partis politiques. In *africavision.info*, 22/01/2013. http://www.africavision.info/

index.php?id=350&tx_ttnews%5Bpointer%5D=28&tx_ttnews%5Btt_news%5D=6223&tx_ttnews%5BbackPid%5D=349&cHash=acf848acf7, checked on 10/08/2014.

Diallo, Ibrahima Talibé. 2013c. Fria entre rêves et réalités. In *guineemining.info*, 24/02/2013. http://www.guineemining.info/index.php?id=285&tx_ttnews%5Btt_news%5D=6509&tx_ttnews%5BbackPid%5D=283&cHash=8a9c2246e6, checked on 11/08/2014.

Diallo, Ibrahima Talibé. 2013d. Fria? In *guineemining.info*, 26/02/2013. http://sanakou.press-guinee.com/index.php?id=285&tx_ttnews%5Bpointer%5D=4&tx_ttnews%5Btt_news%5D=6550&tx_ttnews%5BbackPid%5D=283&cHash=5d46bda85d, checked on 11/08/2014.

Diallo, Ibrahima Talibé. 2013e. Bilan exhaustif de Rusal à Fria. In *guineemining.info*, 4/03/2013. http://www.guineemining.info/index.php?id=285&tx_ttnews%5Btt_news%5D=6640&tx_ttnews%5BbackPid%5D=283&cHash=a6b97087e0, checked on 11/08/2014.

Diallo, Ibrahima Talibé. 2013f. Fria le 15 mars 2013 Suite—Bilan Rusal. In *guineemining.info*, 16/03/2013. http://www.guineemining.info/index.php?id=285&tx_ttnews%5Btt_news%5D=6796&tx_ttnews%5BbackPid%5D=283&cHash=8a6e0ff1a4, checked on 11/08/2014.

Diallo, Ibrahima Talibé. 2013g. Législatif? Exécutifs ont cèdes Dian Dian, et après! Fria c'est quoi? In *guineemining.info*, 27/03/2013. http://www.guineemining.info/index.php?id=285&tx_ttnews%5Btt_news%5D=6969&tx_ttnews%5BbackPid%5D=283&cHash=2b6ca10735, checked on 11/08/2014.

Diallo, Ibrahima Talibé. 2013h. Fria: l'Etat préfèrerait-il Rusal? In *guineemining.info*, 10/04/2013. http://sanakou.press-guinee.com/index.php?id=285&tx_ttnews%5Btt_news%5D=7094&tx_ttnews%5BbackPid%5D=283&cHash=93a6089277, checked on 11/08/2014.

Diallo, Ibrahima Talibé. 2013i. Fria: une solution mais pas la solution. In *guineemining.info*, 15/04/2013. http://sanakou.press-guinee.com/index.php?id=285&tx_ttnews%5Btt_news%5D=7156&tx_ttnews%5BbackPid%5D=283&cHash=4177c60897, checked on 11/08/2014.

Diallo, Ibrahima Talibé. 2013j. Rusal le fléau: Un exhaustif des faits de la vie qui nous emmène à faire l'humanitaire. In *guineemining.info*, 2/08/2013. http://sanakou.press-guinee.com/index.php?id=285&tx_ttnews%5Btt_news%5D=8446&tx_ttnews%5BbackPid%5D=283&cHash=0511bade21, checked on 12/08/2014.

Diallo, Ibrahima Talibé. 2013k. Si Fria m'était conte. In *africavision.info*, 11/08/2013. http://www.africavision.info/index.php?id=350&tx_ttnews%5Bpointer%5D=2&tx_ttnews%5Btt_news%5D=8536&tx_ttnews%5BbackPid%5D=349&cHash=d98f5dd84d, checked on 12/08/2014.

Diallo, Ibrahima Talibé. 2013l. Fria: après les législatives, la misère sera le lot gagnant gagnant a Fria. In *guineemining.info*, 19/10/2013. http://www.guineemining.info/index.php?id=285&tx_ttnews%5Btt_news%5D=9123&tx_ttnews%5BbackPid%5D=283&cHash=4abbc243fa, checked on 13/08/2014.
Diallo, Ibrahima Talibé. 2013m. Fria: Une usine, un peuple, un destin. In *guineemining.info*, 3/12/2013. http://sanakou.press-guinee.com/index.php?id=285&tx_ttnews%5Btt_news%5D=9603&tx_ttnews%5BbackPid%5D=283&cHash=7b446d3981, checked on 13/08/2014.
Diallo, Ibrahima Talibé. 2013n. International Mining Développement: Fria est sauvée! In *guinee7sur7.info*, 19/02/13. http://www.aujourdhui-en-guinee.com/fichiers/videos6.php?code=calb7177&langue=fr&champ2=&pseudo=art14&PHPSESSID=ac6538cb52fccd9b9a803050578780d6, checked on 5/08/2014.
Diallo, El Mohamed Koula. 2014a. Kerfalla Yansané: "Après des décennies de frustration, les mines doivent cesser d'être une malédiction" (discours intégral). In *guinee7.com*, 31/01/2014. http://www.guinee7.com/2014/01/31/kerfalla-yansane-apres-decennies-frustration-les-mines-doivent-cesser-detre-malediction-discours-integral/, checked on 6/08/2014.
Diallo, Ibrahima Talibé. 2014b. Rusal-Firguia: Assemmblée nationale democratique du 13 janvier est-ce une victoire du droit ou la défaite de Rusal? In *guineemining.info*, 3/01/2014. http://www.guineemining.info/index.php?id=285&tx_ttnews%5Bpointer%5D=17&tx_ttnews%5Btt_news%5D=9856&tx_ttnews%5BbackPid%5D=283&cHash=28514e1c0c, checked on 13/08/2014.
Diallo, Ibrahima Talibé. 2014c. Les espoirs vont-ils se concrétiser? In *africavision.info*, 20/01/2014. http://www.africavision.info/index.php?id=350&tx_ttnews[pointer]=15&tx_ttnews[tt_news]=10005&tx_ttnews[backPid]=349&cHash=45d131ae5d, checked on 5/08/2014.
Diallo, Ibrahima Talibé. 2014d. Rusal- Fria: Le DGA de Rusal Guinée annonce la procédure d'acquisition des 50% des salaires annoncé récemment le chef de l'Etat guinéen. In *guineemining.info*, 31/01/2014. http://sanakou.press-guinee.com/index.php?id=285&tx_ttnews%5Bpointer%5D=5&tx_ttnews%5Btt_news%5D=10076&tx_ttnews%5BbackPid%5D=283&cHash=0937135842, checked on 4/08/2014.
Diallo, Ibrahima Talibé. 2014e. Le gouvernement, le patronat, les syndicats - Souvenirs de l'accord tripartite de juin 2006 et les illustres visiteurs. In *guineemining.info*, 13/02/2014. http://www.guineemining.info/index.php?id=285&tx_ttnews%5BpS%5D=1341093600&tx_ttnews%5BpL%5D=2678399&tx_ttnews%5Barc%5D=1&tx_ttnews%5Bpointer%5D=12&tx_ttnews%5Btt_news%5D=10182&tx_ttnews%5BbackPid%5D=283&cHash=1b89f5d266, checked on 6/08/2014.

Diallo, Ibrahima Talibé. 2014f. Suite to Johannes Knierzinger. Fria, 16/02/2014. E-mail.

Diallo, Ibrahima Talibé. 2014g. Suite d'un bon dimanche to Johannes Knierzinger, 21/02/2014. E-mail.

Diallo, Ibrahima Talibé. 2014h. Rusal-Friguia: Abandonnée en plein territoire national, se ronge dans un crime contre l'humanité. In *guineemining.info*, 5/03/2014. http://sanakou.press-guinee.com/index.php?id=285&tx_ttnews%5Bpointer%5D=5&tx_ttnews%5Btt_news%5D=10204&tx_ttnews%5BbackPid%5D=283&cHash=efb5e28031, checked on 12/08/2014.

Diallo, Ibrahima Talibé. 2014i. Rusal-Friguia: les accords de paris les dits retraités, les employés et l'usine, de Rusal-Friguia doivent-ils espérer? In *guineemining.info*, 17/03/2014. http://sanakou.press-guinee.com/index.php?id=285&tx_ttnews%5Bpointer%5D=7&tx_ttnews%5Btt_news%5D=10320&tx_ttnews%5BbackPid%5D=283&cHash=2995ae633b, checked on 13/08/2014.

Diallo, Ibrahima Talibé. 2014j. Mines: Rusal ne reviendra jamais à Fria si ce n'est que pour transporter de la bauxite. In *guineemining.info*, 26/03/2014. http://sanakou.press-guinee.com/index.php?id=285&tx_ttnews%5Btt_news%5D=10439&tx_ttnews%5BbackPid%5D=283&cHash=929bbd62ff, checked on 14/08/2014.

Diallo, Ibrahima Talibé. 2014k. Mines: au secours de Fria. In *guineemining.info*, 11/04/2014. http://sanakou.press-guinee.com/index.php?id=285&tx_ttnews%5Btt_news%5D=10599&tx_ttnews%5BbackPid%5D=283&cHash=34910ead66, checked on 6/08/2014.

Diallo, Ibrahima Talibé. 2014l. Fria: usine et ville sont-elles perdues? In *guineemining.info*, 17/04/2014. http://sanakou.press-guinee.com/index.php?id=285&tx_ttnews%5Btt_news%5D=10666&tx_ttnews%5BbackPid%5D=283&cHash=f9c330fdec, checked on 14/08/2014.

Diallo, Ibrahima Talibé. 2014m. Au fil des dates, Fria sombre doucement et surement!!! In *guineemining.info*, 14/05/2014. http://www.guineemining.info/index.php?id=285&tx_ttnews%5Btt_news%5D=10742&tx_ttnews%5BbackPid%5D=283&cHash=51b740b71c, checked on 5/08/2014.

Diallo, Ibrahima Talibé. 2014n. Bonjour de Fria. E-mail to Johannes Knierzinger. Fria, 8/07/2014. Data file.

Diallo, Ibrahima Talibé. 2014o. Fria est au comble du secours. E-mail to Johannes Knierzinger. Fria, 13/07/2014. Data file.

Diallo, Ibrahima Talibé. 2014p. Rusal annonce l'exploitation de la mine de Dian-Dian en 2016. In *guineemining.info*, 19/07/2014. http://sanakou.press-guinee.com/index.php?id=285&tx_ttnews%5Btt_news%5D=10966&tx_ttnews%5BbackPid%5D=283&cHash=f3d38e1c89, checked on 13/08/2014.

Diallo, Ibrahima Talibé. 2014q. No subject. E-mail to Johannes Knierzinger. Fria, 11/08/2014. Data file.

Diallo, Ibrahima Talibé. 2014r. Fria est au comble du besoin de votre avis. E-mail to Johannes Knierzinger. Fria, 16/08/2014. Data file.

Diallo, Ibrahima Talibé. 2014s. Suite de la veillée avec Fria. E-mail to Johannes Knierzinger. Fria, 21/08/2014. Data file.

Diallo, Ibrahima Talibé. 2014t. Suite de la veillée avec Fria. E-mail to Johannes Knierzinger. Fria, 22/08/2014. Data file.

Diallo, Mouctar. 2014u. Les changements, les permutations et les remerciements attendus. *L'Observateur* 13 (01/2014): 2.

Diallo, Safiatou. 2014v. Massacre de Zogota: l'Ong Même Droit Pour Tous commémore le 2ème anniversaire. In *lejourguinee.com*, 5/08/2014. http://www.lejourguinee.com/index.php/fr/societe/3845-massacre-de-zow-ota-l-ong-droit-pour-tous-commemore-le-2eme-anniversaire, checked on 29/01/2015.

Diallo, Ibrahima Talibé. 2014w. Fria a surement sombre. E-mail to Johannes Knierzinger. Fria, 11/06/2014. Data File.

Diallo, Ibrahima Talibé. 2014x. Le désespoir persiste à Fria. E-mail to Johannes Knierzinger. Fria, 26/02/2014. Data File.

Diallo, Ibrahima Talibé. 2014y. Re: Fria!E-mail to Johannes Knierzinger. Fria, 14/11/2014.

Diallo, Ibrahima Talibé. 2014z. Victoire de Rusal? E-mail to Johannes Knierzinger. Fria, 21/07/2014. Data file.

Diallo, Ibrahima Talibé. 2014aa. Suite d'un bon dimanche to Johannes Knierzinger, 16/02/2014. E-mail.

Diallo, Ibrahima Talibé. 2014ab. Suite et fin des souvenirs to Johannes Knierzinger. Fria, 21/02/2014. E-mail.

Diallo, Boubacar. 2015a. Crise à Fria: L'ancien Directeur de l'usine El hadj Lancei Traoré se prononce! In *Africaguinee.com*. http://www.africaguinee.com/articles/2014/06/02/crise-fria-l-ancien-directeur-de-l-usine-el-hadj-lancei-traore-se-prononce, checked on 28/01/2015.

Diallo, Ibrahima Telly. 2015b. Crise a Fria: Le Ministre des Affaires Etrangères Russes attendu en Guinée. http://www.africatoday.eu/fichiers/videos2.php?langue=fr&idc=fr_CRISE_A_FRIA___Le_Ministre_des_Affaires_Etrang_res_Russes_att, http://www.africatoday.eu/fichiers/livre12-999.php?langue=fr&type=rub36&code=calb5826, checked on 22/03/2015.

Diawara, Ousmane. 2014. *Développement des infrastructures auxiliaires de transport minier en Guinée Conakry*. Conakry: ANAIM.

Die Presse. 2014. Deripaska hält wieder Strabag-Sperrminorität. In *DiePresse.com*, 15/07/2014. http://diepresse.com/home/wirtschaft/economist/3838951/Deripaska-haelt-wieder-StrabagSperrminoritaet-?from=gl.home_wirtschaft, checked on 15/07/2014.

Dulucq, Sophie, Goerg, Odile, and Hélène d', Almeida-Topor. 1989. Les Investissements publics dans les villes africaines, 1930–1985. Habitat et transports. Paris: Editions L'Harmattan (Collection "Villes et entreprises").

Dux, Jörg. 2000. *Guinea—Raumpotential und wirtschaftliche Entwicklungsmöglichkeiten*. Dissertation. Universität Hamburg, Hamburg.

EITI. 2015. Guinea. http://eiti.org/Guinea/.

Emejo, James. 2014. BFI Group: We Never Had Chance to Pay for ALSCON. In thisdaylive.com, 13/04/2014. http://www.thisdaylive.com/articles/bfi-group-we-never-had-chance-to-pay-for-alscon/176023/, checked on 13/08/2014.

Encyclopedia Britannica. 2013. Edea (Cameroon). http://www.britannica.com/EBchecked/topic/178946/Edea, checked on 20/05/2013.

Femmes de Fria. 2013. La lettre ouverte des femmes to Alpha Condé, 8/04/2013. Data file.

Fofana, Aboubacar. 2011. Banankoro: Comment Batax Bouna International mining corporation a miné l'ex Aredor. In *prospectives.over-blog.net*, 31/07/2011. http://prospectives.over-blog.net/article-banankoro-comment-batax-bouna-international-mining-corporation-a-mine-l-ex-aredor-80537572.html, checked on 11/10/2013.

Fommervault, Quitterie de, and Gérard, Vindt. 2001. Paternalisme contre mobilité ouvrière. In *Alternatives Économiques* 189: 62–63.

Forster, Meinhard. 1976. *Struktur und Risiken der deutschen Nichteisen-Metallversorgung. dargestellt am Beispiel Kupfer, Aluminium, Nickel, Zink, Blei und Zinn*. Hamburg: Weltarchiv.

Fortescue, Stephen. 2006. The Russian Aluminum Industry in Transition. *Eurasian Geography and Economics* 47 (1): 76–94.

Gapminder. 2015. Unveiling the beauty of statistics for a fact based world view. http://www.gapminder.org/, checked on 5/02/2015.

Grilli, Enzo R. 1994. *The European Community and the Developing Countries*. Cambridge: Cambridge University Press.

Guinée Emergence. 2014. Mines: A Débélé, K. Yansané salue les efforts de Rusal. In *guineeemergence.com*, 6/10/2014. http://guineeemergence.com/?p=1578, checked on 26/02/2015.

Guinéeinformation. 2014. Interview du professeur Alpha Condé à Genève, sur Rio Tinto, Vale, Simandou, Rusal, Tony Blair, BSGR de Beny Steinmetz, Georges Soros. http://guineeinformation.fr/index.php/guinee/item/753-interview-du-professeur-alpha-conde-a-geneve-sur-rio-tinto-vale-simandou-rusal-tony-blair-bsgr-de-beny-steinmetz/753-interview-du-professeur-alpha-conde-a-geneve-sur-rio-tinto-vale-simandou-rusal-tony-blair-bsgr-de-beny-steinmetz, updated on 30/04/2014, checked on 14/08/2014.

Guinéenews. 2015. Grands dossiers: Vérités et contre-vérités sur le « bradage » de Friguia à 22 millions USD à Rusal. http://www.guineetoile.

com/fichiers/livre12.php?langue=fr&pseudo=&code=calb2635&num=967, checked on 24/06/2015.

Habig, Gutmann. 1983. *Möglichkeiten und Grenzen einer Kontrolle internationaler Rohstoffmärkte durch Entwicklungsländer. Das Beispiel des Kupfer- und Aluminiummarktes*. Hamburg: Weltarchiv.

Haefliger, Markus M. 2012. Wie Guinea Entwicklungschancen verspielt. In *NZZ Online*, 10/01/2012. http://www.nzz.ch/nachrichten/politik/international/wie_guinea_entwicklungschancen_verspielt_1.3880138.html, checked on 10/01/2012.

Harden, Sonya. 2014. FW: Alcoa.com contact email to Johannes Knierzinger, 17/11/2014. E-mail.

Harding, Luke, and Terry, Macalister. 2010. Twilight of the oligarchs as credit crisis hits Russia. In *theguardian.com*, 25/10/2010. http://www.theguardian.com/world/2008/oct/25/russian-oligarchs-oleg-deripaska, checked on 30/06/2014.

Helmer, John. 2005. Ukrainians aim at Deripaska. In *russiajournal.com*, 19/06/2005. http://russiajournal.com/node/19628, checked on 9/08/2014.

Holloway, Steven Kendall. 1988. *The Aluminium Multinationals and the Bauxite Cartel*. New York: St. Martin's Press.

Husband, Charles, McMahon, Gary, and Peter, van der Veen. 2009. The Aluminum Industry in West and Central Africa. Lessons Learned and Prospects for the Future. Extractive Industries and Development Series #13. Washington, DC: World Bank.

Hütz-Adams, Friedel, Voge, Ann-Kathrin, Liese, Sascha, Hörmann, Stefan, and Jan Per, Bethge. 2014. Nachhaltige Rohstoffe für den deutschen Automobilsektor. Herausforderungen und Lösungswege. Bonn: Südwind e.V.

Huy, Dieter. 2013. *Deutschland—Rohstoffsituation 2012*. Hannover: Bundesanstalt für Geowissenschaften und Rohstoffe.

IMF. 2014. International Monetary Fund Home Page. http://www.imf.org, updated on 17/12/2014, checked on 18/12/2014.

International Monetary Fund. 2008. Guinea: Selected Issues and Statistical Appendix. IMF Country Report 08/20. Washington, DC, checked on 5/09/2014.

International Monetary Fund. 2012a. Guinea: Poverty Reduction Strategy Paper—Annual Progress Report. Washington, DC.: IMF Publication Services. http://www.imf.org/external/pubs/ft/scr/2012/cr1261.pdf, checked on 5/11/2013.

International Monetary Fund. 2012b. Press Release: IMF and World Bank Announce US$2.1 Billion Debt Relief for Guinea. https://www.imf.org/external/np/sec/pr/2012/pr12363.htm, updated on 26/09/2012, checked on 15/01/2015.

Investguinée. 2012. Le patron de Rusal Oleg Déripaska. In investguinee.com, 14/04/2012. http://www.investguinee.com/portrait/le-patron-de-rusal-oleg-deripaska/47, checked on 12/08/2014.

Investir au Cameroun. 2014. Rio Tinto explique les raisons de son retrait d'Alucam, le géant de l'aluminium au Cameroun. In *investiraucameroun.com*, 20/10/2014. http://www.investiraucameroun.com/entreprises/2010-5750-rio-tinto-explique-les-raisons-de-son-retrait-d-alucam-le-geant-de-l-aluminium-au-cameroun, checked on 9/01/2015.

Jeune Afrique. 2009. Quand le « Dadis show » dégénère. In *Jeuneafrique.com*, 14/04/2009. http://www.jeuneafrique.com/Article/ARTJAJA2518p008-009.xml3/usine-moussa-dadis-camara-anatoly-pantchenko-ruski-aluminiquand-le-dadis-show-degenere.html, checked on 23/10/2014.

Johnson, R.W. 1970. Sékou Touré and the Guinean Revolution. *African Affaires* 69 (277): 350–365.

Johnson-Sahr, Eric. 1991. *Livret de Stagiare. Apprenons a connaitre notre milieu de travail. Centre de Developpement et de Formation*. Kamsar: CBG.

Jolly, Thomas. 2015. FW: AW: Destination von Bauxit aus Guinea to Johannes Knierzinger. Lausanne, 13/01/2015.

Kaba, Lansine. 1977. Guinean Politics: A Critical Historical Overview. *The Journal of Modern African Studies* 15 (1): 25–45.

Kadyrov, Raphael. 2010. Bilan du mois de mai par rapport aux 5 premiers mois, performance et défis. *La voix de Rusal* 40 (2): 3.

Kadyrov, Raphael. 2011. Proces Verbal de negotiations tenant lieu de protocol d'accord. Direction Génerale, Usine Kimbo, Préfecture de Fria, 15/09/2011.

Kennedy, Dominic. 2008. To survive the aluminium wars, mettle was needed. In The Times, 5/07/2008.

Khrennikov, Ilya. 2012. Rusal Chairman Quits Citing 'Crisis' at Aluminum Producer. In *bloomberg.com*, 13/03/2012. http://www.bloomberg.com/news/2012-03-13/rusal-chairman-quits-citing-crisis-at-aluminum-producer.html, checked on 7/08/2014.

Knierzinger, Johannes. 2014. The Socio-Political Implications of Bauxite Mining in Guinea: A Commodity Chain Perspective. *The Extractive Industries and Society* 1: 20–27.

Knierzinger, Johannes. 2016. Bauxit und Aluminium aus Afrika: Ausbeutung auf Umwegen. In *Globale Ressourcen und Rohstoffpolitik. Aktuelle Auseinandersetzungen im historischen Kontext. Historische Sozialkunde/Internationale Entwicklung 35*, ed. Karin Fischer, Johannes Jäger, and Lukas Schmidt, pp. 203–220. Wien: New Academic Press.

Knight Piésold, and Co. 2008. Guinea Alumina Project. Social and Environmental Assessment. Volume 1—Executive Summary. Denver: Guinea Alumina Corporation. http://www.guineaalumina.com/assets/files/Volume1/Executive%20Summary%20Text.pdf, checked on 11/10/2013.

Lanning, Greg, and Marti Mueller. 1979. *Africa undermined. Mining Companies and the Underdevelopment of Africa*. Penguin (A pelican original): Harmondsworth.

Larrue, Jacques. 1991. Fria en Guinée. Des aspects humains d'une industrialisation differente. *Cahiers d'histoire de l'aluminium* 7: 37–49.

Larrue, Jacques. 1997. *Fria en Guinée. Des aspects humains d'une industrialisation differente*. Karthala: Paris.

Le Figaro. 2008. Le Russe qui a perdu 20 milliards d'euros. In *lefigaro.fr*, 26/12/2008. http://www.lefigaro.fr/patrimoine/2008/12/24/05001-20081224ARTFIG00242-la-crise-coute-milliards-a-oligarques-russes-.php, checked on 29/04/2015.

Le Figaro. 2011. Rusal proteste contre l'Etat guinéen. In *lefigaro.fr*, 18/02/2011. http://www.lefigaro.fr/flash-eco/2011/02/18/97002-20110218FILWWW00453-rusal-proteste-contre-l-etat-guineen.php, checked on 5/08/2014.

Le Guinéen. 2013. 92% des travailleurs demandent pardon à Rusal. In *leguineen.info*, 24/04/2013. http://www.leguineen.info/articles/economie-soci-t/92-des-travailleurs-demandent-pardon-rusal, checked on 29/10/2014.

Leimgruber, Walter. 1990. *Kalter Krieg um Afrika*. Stuttgart: Franz Steiner Verlag.

Leno, Fara Nestor. 2014a. *Etudes Sommaires de la Monographie de Sangarédi*. Sangarédi: Service Développement Rural de Sangarédi.

Leno, Fara Nestor. 2014b. Règles essentielles pour un dédommagement agricole. Service Développement Rural de Sangaredi.

L'Observateur. 2014. Des nouvelles locomotives pour Rusal-CBK. In L'Observateur, 20/01/2014 (682).

London Metal Exchange Limited. 2014. London Metal Exchange: Aluminium. London Metal Exchange Limited. http://www.lme.com/metals/non-ferrous/aluminium/#tab2, updated on 24/10/2014, checked on 24/10/2014.

Manson, Katrina, and James, Knight. 2007. Guinea takes control of its minerals. In *BBC News*, 15/01/2007. http://news.bbc.co.uk/2/hi/business/6254719.stm, checked on 2/02/2012.

Marshall, Svetlana. 2013. Hururu protest threatens $3.4B investment. In *Guyana Times*, 26/04/2013. http://www.guyanatimesgy.com/2013/04/26/hururu-protest-threatens-3-4b-investment-rusal/, checked on 15/10/2014.

Martin, Guy. 1982. Africa and the Ideology of Eurafrica: Neo-Colonialism or Pan-Africanism. *The Journal of Modern African Studies* 20 (2): 221–238.

Marx, Adolf. 1977. Verflucht, wer uns vergißt. Gefoltert für Deutschland. Much: Derscheider (5).

MBendi Information Services. 2013a. Gold Mining in Guinea. http://www.mbendi.com/indy/ming/gold/af/gu/p0005.htm, updated on 10/10/2013, checked on 11/10/2013.
MBendi Information Services. 2013b. Diamond Mining in Guinea. http://www.mbendi.com/indy/ming/dmnd/af/gu/p0005.htm, updated on 11/10/2013, checked on 11/10/2013.
Metal Bulletin. 2013. Nigeria pushes ahead with Alscon sale to BFI; Rusal contests. In metalbulletin.com, 31/01/2013. http://www.metalbulletin.com/Article/3149323/Nigeria-pushes-ahead-with-Alscon-sale-to-BFI-Rusal-contests.html, checked on 13/08/2014.
Moore Stephens. 2013. *Rapports de Réconciliation ITIE Guinée. Années 2011&2012.* Conakry: EITI.
Morgaut, Marc-Edmond. 1960. La Promotion Sociale en Afrique. *In Bulletin Alucam* 13: 13–14.
N'Diaye, Aboubacar. 1984. Les mutations économiques en Basse Guinée de 1958–1984. Mémoire de diplôme de fin d'études supérieures, Conakry.
Neue Zürcher Zeitung. 2010. Rusal legt schwaches Börsendebüt in Hongkong hin. In NZZ.ch, 27/01/2010. http://www.nzz.ch/nachrichten/rusal-legt-schwaches-boersendebuet-in-hongkong-hin-1.4656831, checked on 5/08/2014.
Nodalis, Sedec, mtbs. 2013. *Développement des infrastructures auxiliaires de transport minier en Guinée Conakry.* ANAIM: Rapport de diagnostic—previsoire. Conakry.
Nouvelles de Guinée. 2014. Fria: des citoyens en colère incendient les locaux de la gendarmerie à la suite de la mort d'un jeune. In nouvelledeguinee.com, 26/02/2014. http://nouvelledeguinee.com/fichiers/blog16b.php?type=rub2&langue=fr&code=calb737.
OANDA. 2015. Trading du Forex. http://www.oanda.com/lang/fr/, checked on 1/04/2015.
Odey, Patrick. 2013. Nigeria: Alscon—a Tale of Vainly Privatised Smelter. In *allAfrica.com*, 17/10/2013. http://allafrica.com/stories/201310170330.html, checked on 15/10/2014.
OEC. 2015. The Observatory of Economic Complexity. https://atlas.media.mit.edu/en/, checked on 23/09/2015.
Oxfam. 2015. Global Food Index. http://www.oxfam.org.uk/what-we-do/good-enough-to-eat, checked on 17/02/2015.
Pauthier, Céline. 2002. *Fria: Une ville-usine en Guinée. Des dernières années de la colonisation à la République Populaire 1956–1973. Mémoire de maitrise d'histoire sous la direction d'Odile Goerg.* Paris: Université Denis Diderot Paris VII.
Pechiney Ugine Kuhlmann. 1976. Le complexe d'aluminium en U.R.S.S.: une première étape est franchie. In Jalons 12, p. 7.

Pechiney Ugine Kuhlmann. 1977. La guenon, la locomotive et la bauxite. In Aluminium Pechiney 31.

Pechiney Ugine Kuhlmann. 1982. Edéa: un cadre de vie agréable. *In Jalons* 32: 4.

Permatasari, Soraya. 2012. Alcoa to Consider Buying BHP Billiton, Rio Tinto Aluminum Assets. In *bloomberg.com*, 21/11/2012. http://www.bloomberg.com/news/articles/2012-11-21/alcoa-to-consider-buying-bhp-billiton-rio-tinto-aluminum-assets, checked on 13/09/2013.

Pezet, Anne, Loison, Marie-Claire, and Céline, Berrier. 2009. Corporate Social Responsibility before CSR. Practices at the Company Aluminium du Cameroun, 1950s–1970s.

Préfecture de Fria. 23/10/2009. Procès-verbal tenant lieu de protocole d'accord.

Radio Netherlands Worldwide. 2012. Guinean town paralysed as pay strike shuts aluminium plant. In *Radio Netherlands Worldwide Africa*, 26/08/2012. http://www.rnw.nl/africa/bulletin/guinea-town-paralysed-pay-strike-shuts-aluminium-plant, checked on 24/05/2013.

Ralph Spencer Associates. 2000. *Co-operation in the Mining Sector and SYSMIN. Evaluation Synthesis.* Brussels: Evaluation Unit (SCR/F/5) of the European Commission.

République de Guinée. 2006. Annex C de la convention du 03 novembre entre la République de Guinée et "Rousski Alumini" pour la réhabilitation, l'extension et l'exploitation des gisements de bauxite de Kindia. Conakry.

République de Guinée. 2011. Annuaire gestion des hôpitaux publics.

République de Guinée. 2013a. Code Minier 2011 amendé. Conakry, updated on 25/10/2011, checked on 16/05/2014.

République de Guinée. 2013b. Comité Technique de Revue des Titres et Conventions Miniers. http://www.contratsminiersguinee.org/about/projets.html, updated on 17/08/2013, checked on 4/10/2013.

République de Guinée. 2013. *Compagnie des Bauxites de Guinée.* Conakry: Historique en bref.

République de Guinée. 2013d. Titres Miniers Pour La Bauxite. Situation au décembre 2012.

République de Guinée. 2014. Arrêté N° 3015/MP/SG/2015. Portant publication des résultats préliminaires du troisième recensement général de la population et de l'habitation réalisé du 1er Mars au 02 Avril 2014. http://www.stat-guinee.org/index.php/result-prelim-rgph3?download=8:res-pre-rgph3, checked on 27/01/2015.

Revoin, Jean-Didier. 2012. Rusal: quand le politique l'emporte sur la finance. In *lecourrierderussie.com*, 21/06/2012. http://www.lecourrierderussie.com/2012/06/rusal-quand-le-politique-l-emporte-sur-la-finance/, checked on 8/08/2014.

Ribadeau-Dumas, Jacques. 1952. Parlez-nous de l'Afrique. *In Le Bulletin Pechiney* 55: 17–26.

Rio Tinto Alcan. 2015. The global leader in the aluminum industry. http://www.riotintoalcan.com/, checked on 9/01/2015.

Riverton, John. 2014. The Volta River Project. In *Global Poverty Alleviation: A Case Book*, ed. P.J. Albert, et al., 31–56. Heidelberg: Springer.

Rusal. 2010a. La Cité de Débélé a désormais une pharmacie. *In La voix de Rusal* 40 (2): 8.

Rusal. 2010b. RUSAL/CBK: 40,000 dollars pour 3 forages dans la cité de Simbaya. *In La voix de Rusal* 40 (2): 4.

Rusal. 2011a. Le prix de revente des usines d'alumine de Rusal. *In La voix de Rusal* 57 (12): 4.

Rusal. 2011b. Pose de la première pierre du Centre de santé de Fomah. *In La voix de Rusal* 51 (6): 11.

Rusal. 2011c. RUSAL/FRIGUIA: Un lot de médicaments pour soulager les agents et leurs familles. *In La voix de Rusal* 56 (11): 4.

Rusal. 2012. BHP Billiton renonce à son projet de transformation de la bauxite en alumine de Boffa. *In La voix de Rusal* 58: 5.

Rusal. 2014a. Compagnie des Bauxites de Kindia (CBK). http://www.rusal.ru/en/about/34.aspx, checked on 24/09/2014.

Rusal. 2014b. Facts and Figures. http://www.rusal.ru/en/about/facts.aspx, updated on 15/10/2014, checked on 15/10/2014.

Rusal. 2014c. Le Marché de l'aluminium toujours déséquilibré. In *La voix de Rusal* 76 (12).

Rusal. 2014d. Programme Social de Rusal CBK en 2013. *In La voix de Rusal* 76 (12): 10.

Rusal. 2014e. Rusal continue la réduction des capacités non efficaces. In *La voix de Rusal* 76: 2.

Rusal. 2014f. Sayanal. http://www.rusal.ru/en/about/13.aspx, checked on 17/10/2014.

Saada, Assane. 2009. Le Dadis Web Show. In *Jeuneafrique.com*, 9/09/2009. http://www.jeuneafrique.com/Article/ARTJAJA2539p036-037.xml0/, checked on 14/01/2015.

Sabaté, Jean. 2009. Le syndrome Pikalevo. In Le Monde diplomatique, 12/2009. http://www.monde-diplomatique.fr/2009/12/SABATE/18634, checked on 14/01/2015.

Samb, Saliou. 2006. The Electricity Puzzle. *In La Revue—Magazine semestriel du bureau de la banque mondiale en Guinée* 1: 18–21.

Schicho, Walter. 2001. *Handbuch Afrika. Band 2*, vol. 2. Frankfurt am Main: Brandes & Aspel.

Schmidt, Elisabeth. 2007. *Cold War and Decolonization in Guinea, 1946–1958.* Athens: Ohio University Press.

Schmitter, Philippe C., and Gerhard Lehmbruch (eds.). 1979. *Trends Towards Corporatist Intermediation*. Beverly Hills: SAGE.

Senghaas-Knobloch, Eva. 1975. The Internationalization of Capital and the Process of Underdevelopment. The Case of Black Africa. *In Journal of Peace Research* 12 (4).

Shafer, D. Michael. 1986. Undermined: The Implications of Mineral Export Dependence for State Formation in Africa. *Third World Quarterly* 8 (3): 916–952.

Shundeyev, Vladimir. 1974. The Republic of Guinea. *In The African Communist* 58: 34–45.Sidibé, Alhassane. 2012. Zoom sur l'usine. 2eme partie reprise (24). Voix de Fria, 27/05/2012.

Sidibé, Alhassane. 2012a. Zoom sur l'usine. 2 Milliards (25). Voix de Fria, 27/05/2012.

Sidibé, Alhassane. 2012b. Zoom sur l'usine. Payement des salaires (28). Voix de Fria, 27/05/2012.

Sidibé, Alhassane. 2012c. Zoom sur l'usine. Condition (27). Voix de Fria, 22/07/2012.

Sidibé, Alhassane. 2012d. Zoom sur l'usine. Levée de grève (27). Voix de Fria, 15/07/2012.

Sidibé, Alhassane. 2012e. Zoom sur l'usine. Cellou Dalein (19). Voix de Fria, 8/08/2012.

Sidibé, Alhassane. 2012f. Zoom sur l'usine. Payement de salaire (13). Voix de Fria, 22/08/2012.

Sidibé, Alhassane. 2012g. Zoom sur l'usine. Sages et travailleurs (28). Voix de Fria, 6/10/2012.

Sidibé, Alhassane. 2012h. Zoom sur l'usine. Débrayage des postes (20). Voix de Fria, 21/10/2012.

Sidibé, Alhassane. 2012i. Zoom sur l'usine. Juge (21). Voix de Fria, 17/11/2012.

Sidibé, Alhassane. 2013. Zoom sur l'usine. Reprise annoncée (16). Voix de Fria, 9/03/2013.

Sidibé, Alhassane. 2013a. Zoom sur l'usine. Diagnostic (9). Voix de Fria, 17/03/2013.

Sidibé, Alhassane. 2013b. Zoom sur l'usine. Rétrospective 1 (23). Voix de Fria, 3/04/2013.

Sidibé, Alhassane. 2013c. Zoom sur l'usine. Blocus préfecture (5). Voix de Fria, 17/04/2013.

Sidibé, Alhassane. 2013d. Zoom sur l'usine. Les vols à l'usine (12). Voix de Fria, 29/05/2013.

Sidibé, Alhassane. 2013e. Zoom sur l'usine. Pétition (14). Voix de Fria, 24/04/2013.

Sidibé, Alhassane. 2013f. Zoom sur l'usine. Audit (3). Voix de Fria, 22/05/2013.
Sidibé, Alhassane. 2013g. Zoom sur l'usine. Décision de justice (8). Voix de Fria, 3/07/2013.
Sidibé, Alhassane. 2013h. Zoom sur l'usine. Crise sur le marché (1). Voix de Fria, 10/07/2013.
Sidibé, Alhassane. 2013i. Zoom sur l'usine. Blocage de l'entrée (4). Voix de Fria, 31/07/2013.
Sidibé, Alhassane. 2013j. Zoom sur l'usine. Négociation (22). Voix de Fria, 7/08/2013.
Sidibé, Alhassane. 2013k. Zoom sur l'usine. Réaction des syndicats (15). Voix de Fria, 14/08/2013.
Sidibé, Alhassane. 2013l. Zoom sur l'usine. Incident des femmes (10). Voix de Fria, 4/12/2013.
Sidibé, Alhassane. 2013n. Zoom sur l'usine. Retro: 3 et 4 avril 2012 (17). Voix de Fria, 26/06/2013.
Soloviev, Vladislav. 2011. Nous menons un dialogue avec les nouvelles autorités guinéennes. *In La voix de Rusal* 48 (3): 1–3.
Soumah, Ibrahima. 2008. *The Future of Mining Industry in Guinea*. Paris: L'Harmattan.
Soumah, Momo. 2012. Mahmoud Thiam: Cinglante réplique et grand déballage. In *guineeconakry.info*, 5/12/2012. http://www.guineeconakry.info/article/detail/mahmoud-thiam-cinglante-replique-et-grand-deballage/, checked on 5/08/2014.
Storozhenko, Mikhail. 2009. Aide-mémoire sur la privatisation de Friguia to Vincent Ernens, Boubacar Hann, Mory Kaba, Gaoussou Alhassane Keita, Lamine Sylla, Abdoul Soundiata Konate et al. Conakry, 30/04/2009. E-mail.
Sylla, Kanfory. 2013. Visite du Premier Ministre a Fria. In *guineemining.info*, 11/09/2013. http://sanakou.press-guinee.com/index.php?id=285&tx_ttnews%5Btt_news%5D=8811&tx_ttnews%5BbackPid%5D=283&cHash=99705ebb4c, checked on 13/08/2014.
Tcheby, Eugène. 1960. La page camerounaise: Causeries sur le travail. In Bulletin Alucam 13.
The Express Tribune. 2012. Abu Dhabi Inks Aluminum Ore Supply Deal With Guinea. In *tribune.com*, 3/11/2012. http://tribune.com.pk/story/460498/aluminium-ore-abu-dhabi-inks-supply-deal-with-guinea/, checked on 25/06/2015.
The Moscow Times. 2015. Ukraine Supreme Court Upholds Nationalization of Rusal Aluminum Plant. In *The Moscow Times*, 12/03/2015. http://www.themoscowtimes.com/business/article/ukraine-supreme-court-upholds-nationalization-of-rusal-aluminum-plant/517382.html, checked on 5/05/2016.

Tilouine, Joan. 2014. Guinée: le cauchemar de Fria. In *Jeuneafrique.com*, 24/02/2014. http://economie.jeuneafrique.com/regions/afrique-subsaharienne/21422-guinee-le-cauchemar-de-fria.html#Reaction_rusal, checked on 7/08/2014.
Tilouine, Joan, and Joël, Té-Léssia. 2014. Guinée: Rusal démarre le projet de bauxite de Dian-Dian. In *Jeuneafrique.com*, 11/07/2014. http://economie.jeuneafrique.com/regions/afrique-subsaharienne/22579-guinee-rusal-demarre-le-projet-de-bauxite-de-dian-dian.html, checked on 7/08/2014.
Tourot, Nicolas. 2003. *Trois grands projets de développement hydroélectrique en Afrique Noire, mythes et réalisations: Des mythes coloniaux aux éléphants blancs contemporains.* Mémoire. Paris: Université Panthéon-Sorbonne.
Touré, Alhadji Mohamed Lamine. 2009. Mémoires d'un compagnon de l'independence Guinéenne. Paris: L'Harmattan.
Touré, Ahmed Sékou. 2010. *La lutte pour le vote et l'application du Code du travail d'Outre-Mer à travers l'Ouvrier en 1952–1953.* Conakry: Club Ahmed Sékou Touré.
USAID. 2006. Guinea Operational Plan. Silver Spring, checked on 5/09/2014.
US Department of Labor. 2014. Bureau of Labor Statistics. http://www.bls.gov/home.htm, checked on 15/11/2014.
USADF. 2015. US African Development Foundation—Creating Pathways to Prosperity for Underserved Communities. http://www.adf.gov/, updated on 16/12/2014, checked on 8/01/2015.
USGS. 2016. US Geological Survey. http://minerals.usgs.gov, checked on 15/06/2016.
Uwechue, R. 1991. *Makers of Modern Africa*. London: Africa Books.
Vassiliev, Pavel. 2012. Ensemble combattons la crise. *In La voix de Rusal* 58: 6–8.
Vision Guinée. 2013. Le Salaire minimum en Guinée fixé à 440,000 GNF. In *visionguinee.info*, 30/04/2013. http://www.visionguinee.info/2013/04/30/le-salaire-minimum-en-guinee-fixe-a-440-000-gnf/, checked on 20/11/2014.
Vitry, Raoul de. 1957. Éditorial. In Le Bulletin Pechiney (Christmas issue), pp. 2–3.
Vuerings, Raymond. 1964. Problèmes de gestion d'entreprise en pays sous-développés: La reconversion de l'entreprise privée en Afrique noire indépendante. *Management International* 4 (4): 93–104.
Wiederstein, Alexis. 1994. *Alu-Wende. Eine weltweite ökologische und soziale Bestandsaufnahme der österreichischen Aluminiumwirtschaft und Aspekte zur Neuorientierung.* Wien: Ökologie-Institut.
Windfuhr, Simone. 2004. *Zweifelhafte Beratung: Der Einfluss der Weltbank auf Minengesetze in Afrika.* FIAN: Menschenrechtliche Anmerkungen. Köln.
World Bank. 1990. Republic of Guinea. Country Economic Memorandum. Report No. 8774. Washington, DC.

World Bank. 2015. Inflation, consumer prices (annual %). http://data.worldbank.org/indicator/FP.CPI.TOTL.ZG, updated on 25/02/2015, checked on 26/02/2015.

World Investment News. 2009. Interview avec Monsieur Ibrahima Soumah. http://www.winne.com/topinterviews/soumah.htm, updated on 2/03/2009, checked on 1/03/2012.

CHAPTER 4

Conclusions

4.1 Global Chains of Command

Based upon the tool kit of global production network analysis, I have tried to show in this study that considerable parts of African political decision spaces are dominated by the transnational corporate chains of command, stretching from various headquarters of the TNCs to the caterpillar drivers in the mines, and from there to homemakers, artisans, traders, and peasants who contribute either directly to cheaper consumer products and higher profits or do so by keeping the reproduction costs of the core workforce low. In the case of Guinea, the "drivers" of these multipolar chains (cf. Raikes et al. 2000: 7) are the vertically integrated aluminum companies as well as manufacturers and branded companies (cf. Gereffi 1997: 116f), and all of these firms are strongly entangled with state administrations or are in fact state owned. As the few bauxite miners also provide a large part of the taxes that go into governmental salaries (investment is mostly driven from outside), the whole Guinean state apparatus and thereby the whole Guinean population are part of these global chains of command (cf. with the term "souveraineté partagé" as used by Magrin 2013: 197; see also Larrue 1997: 92). Therefore, social hierarchy in Guinea and particularly in the mining towns strongly depends on transnational corporate decision spaces, mainly aiming at profit maximization, but, nonetheless, leaving ample room for brokers at every tier of the chain: while CEOs decide upon the fate of whole cities, their community relation departments decide

J. Knierzinger, *Bauxite Mining in Africa*, International Political Economy Series, DOI 10.1007/978-3-319-52706-2_4

over the building of schools, wells, and clinics and environmental offices decide over relocations and compensations of peasants near the mines. When the currently about 3500 employees of the bauxite mines go on strike, they threaten to cutoff the state from half of its influx of foreign currency. On every node of such a global chain of command (e.g., from an investment decision in Germany to the relocation of a village in Guinea), actors are thereby always at the same time "power brokers". By passing down orders from above (e.g., the CEO of an enterprise trying to raise the profit for the shareholders), they can always use their position for their own means.

This expanding company hierarchy is accompanied by strongly monetarized local control relations, which overshadow all other previously existing modes of rule in the areas, whether they are based upon elections, ethnic ties, religion, gender, or other conventions. During the development period of the current Guinean mines from the 1950s to the 1970s, large numbers of highly paid technicians and managers from the world economic centers had to be attracted and comforted to stay at least for several years. These bearers of high cultural capital also exchange working time with people on the local markets, buy land, send their kids to school, etc. According to my own calculations, an expatriate geologist buying oranges on the streets of Sangaredi exchanges working time in the ratio of up to thousand to one with local venders: for working one day in the mines, he could tell thousand street hawkers what to do for one day. This high grade of visible unequal exchange can be compared with global trade centers (c.f. Sassen 2005).

During the 1970s and 1980s, most of these expatriates were replaced by what could be called a Guinean "labor aristocracy" (cf. Larmer 2006; Arrighi 2002: 11), which today possesses large parts of the land and owns enterprises and even schools in these towns. All the analyzed cities can be separated in a workers' core zone and a non-worker periphery (cf. Interview with Activist 2/03/2014; Larrue 1997: 236). However, the borders of these islands of prosperity are quite difficult to define, given that many core tasks have been outsourced to badly paid subcontractors in all the concerned cities in the last 30 years (cf. Lee 2009). Most of the security guards, maintenance personnel, builders, and other artisans working in the mines earn less than the Guinean minimum wage of about 60 USD per month. Moreover, as the workers lose both income and fringe benefits with their retirement and as retirement schemes are poorly developed in Guinea, these zones also have a temporal limit, whereby entire extended families are regularly forced out of their houses and plunged into poverty due to the retirement of a single worker.

However, this highly unequal division of the right to command based upon property and income can also be traced back along the value chain and from there to the investors (and in a way also to the consumers). Out of convenience, we take again the concrete example of Oleg Deripaska, the current president of Rusal: with his reported annual wage of 26.5 million USD (in 2012; Investguinée 2012), which means that—without his income from capital—he earns roughly 570 times more than the German GNP per head of 46,700 USD (2012)[1] and 60,000 times more than the Guinean GNP per head of 440 USD, also in 2012 (World Bank 2014). While working 1 h for the global market—for instance, by discussing about the future of Fria with the Guinean president in his palace in Conakry—he would thereby earn enough money to hire 60,000 Guineans to parade for him on the streets to convince the president of his popularity in the country (cf. Knierzinger 2014). This is a level of "unequal exchange" (see ibid.) that can clearly be called "political". In order to minimize taxes, salaries, and other local expenditure, the commanders of the aluminum production network continuously give out presents to pacify the population by building infrastructure, employing or sending away trouble-makers, paying primes to local police forces, etc. Out of this underlying structure of capitalist exploitation and unequal exchange—and thereby unequal relations of control on a global scale—evolve without doubt the most constricting power relations for Guineans. Next to this legalized form of exploitation, corruption scandals—besides Ebola, the only topic in Guinea that seems to hold any interest for international media—are negligible.

Attempts to fundamentally change these very concrete control relations currently do not figure in discussions about development, probably because they would involve rethinking our idea of market economy as such. One exception are discussions around Thomas Piketty's "Capital in the twenty first Century" that also brought to light the mounting income gap between the very rich and the rest in the US and elsewhere (cf. Krugman 2014). From a ratio of about 30:1 in the late 1970s, the CEO-to-worker compensation ratio in the US rose to currently around 500:1 (Mishel and Davis 2014). However, in discussions about African development—given that the concept of development is not rejected as such—these realities do not seem to be of importance. Even the most progressive actors in Guinean bauxite towns seemed to think that the

Extractive Industries Transparency Initiative (EITI) is "all we can get". It is true that the need for more transparency in the global aluminum industry is great: in many cases, taxes do not arrive, where they should and it generally remains questionable whether the operating units of the aluminum industry can be regarded as "companies" governed by "market forces" (cf. Taylor 2000: 7). The main aim of Publish What You Pay and the EITI is nonetheless fighting corruption, which means behavior that is considered unlawful from the start. Compared to the legally justified capital relations resulting from globalized production networks, these crimes are negligible. The vividly debated question concerning whether the Condé government may or may not have taken bribes from the mining majors diverts the attention from the fact that much higher sums are redirected to the capitalist centers through tax evasion (for transfer pricing in the aluminum industry, see Bauer and Maissen 1989, 92–100), unequal exchange, environmental destruction, and capitalist exploitation (Knierzinger 2014). In the case of African bauxite mining and processing, singular physical persons (who happened to be investors and/or CEOs) decide upon the fate of millions by telling their workers to stop the machines or their bank assistants to close an account, without breaking any laws.

The difference of Africa's bauxite mining towns with other places thus lies in the visibility of the international division of labor in these cities (and the consequences of this close interaction of unequal partners for the hierarchical order on-site) and the huge gaps in control relations between global capital—many of the investors and philanthropists who are active in Africa form part of the richest people on the planet—and an unorganized (semi-)proletariat that figures among the lowest ranks of global poverty indices. Besides being subjected to capitalist overexploitation, these mining towns are thereby places, where spatially separated levels of global control emerge from the hidden fabric of world economic exchange and interfere with local systems of rule. Normally, invisible global control relations between consumers of OECD countries and producers in developing countries are shrunk to the size of cities and become part of everyday experience. While these relations are as real in the rest of the world as at these junctures, they are normally mediated through long commodity chains.

In the short and medium terms, this clash of highly unequal cultural capital leads, on the one hand, to dispossession, disempowerment, and exploitation and can lead, on the other hand, to formal employment, better medical care, and generally better living conditions, at least for a fraction of the local population. In the long term, this also implies the mostly unrealized possibility of knowledge transfer and spin-off effects and thus more self-sufficient development that would be less affected by external crises. The latter point appears to be the most important and unsolved to date, corresponding to what could be called the socio-political aspects of the paradox of the plenty (Magrin 2013: 103).

4.2 Different Company Cultures

All mentioned consequences of bauxite mining in Guinea depend to a certain degree on the "business culture" of the companies involved, although this should not be overstated given that most corporate behavior can be explained on the basis of the economic and political conditions under which these companies operate. The history of Fria would probably have been different had it remained within the sphere of influence of Western companies, because highly different forms of state are connected to Rusal and Alcoa. The highly personalized oligarchic capitalism in modern-day Russia results in different types of companies than Western forms of oligopoly capitalism. In addition, following a pattern that can be traced all over Africa, Rusal had taken over Fria during a conflictual situation when European and US companies had already withdrawn. This is important to note because it touches on our understanding of the agency of CEOs and investors. If Rusal had sold Norilsk, the factory of Fria would probably still be running. Considering the rather simple ownership structure of Rusal, one cannot but recognize that Rusal's majority owner Oleg Desipaska bears some responsibility for the social debacle in Fria. However, as the history section on Fria showed, the company only took over the management of decline of a mining town that had begun under Western leadership (cf. Knierzinger 2016: 156).

As a matter of fact, these differences in structural conditions, cultural norms, and prevalent ideologies can also be made out between Western companies. While Pechiney had developed a corporatist and paternalist approach that was strongly influenced by catholic corporatism, the "civilizing mission" of American companies was focused much more on doing business. However, despite the low value added of bauxite mining, all of these privately-run companies—as well as the highly indebted state (Shafer 1986: 918)—follow the dictate of profit maximization that tends to overrule corporate social responsibility measures, particularly in times of crisis. In addition, both joint ventures were forced to accumulate infrastructural power and thereby to develop modes to cope with this dominant position.

The worker's culture in Fria certainly had an important influence on the conflict with Rusal. Beside Pechiney's catholic corporatism (see Schmitter and Lehmbruch 1979) and the strong influence of French Marxism on trade unions, this culture was certainly also strongly marked by Sékou Touré's understanding and praxis of African socialism that rejected chieftaincy while at the same time describing the ideal (African) society in a particularly harmonic and romantic way. According to Johnson, Touré actually opposed the idea of trade unionism as an instrument of class struggle within Guinea and rather saw it as an instrument of anti-colonialist struggle and of "harmonious evolution and rapid emancipation" (Johnson 1970: 356). Despite his rejection of colonialism, many elements of Touré's rule can be compared to the early corporatists of the late nineteenth century, which also highly influenced the company culture of Pechiney (for an attempt to adapt Philippe C. Schmitter's concept of corporatism to Africa, see Nyang'oro 1989). After Touré's purge in Fria in the mid-1970s, the only active trade unionists remained in Kamsar and Sangaredi. They continued to receive contracts and proceeded with negotiations on payment and living conditions during the Touré regime and thereby became spearheads of the national trade union movement after his death (Interview with Trade unionist 17/04/2014).

The entry of the US junior company Global Alumina in the 2000s probably showed how far "company culture" can go. Announcing a "historic turning point for the country" (Committee on foreign affairs 2007: 45), the company introduced unprecedented social and ecological standards in Guinea and created expectations of the emergence of a new "Little Paris" in the sub-prefecture of Sangaredi (the former had been Fria). However, with the end of the boom and the takeover of Arab investors, this company has also left more Guineans disenchanted than any other (Interview with Sociologist GAC 1/02/2014). These moments when the leading companies' change show both the impact of "company cultures" on the local population and workforce and their importance for national politics. A trade union activist in Fria stated that the fundamental problem of the "Russians" in Fria was their "inheritance of a company that has occidental culture." According to him, Débélé never had similar problems, because it had been dominated by "Russian culture" from the start. Similar to Touré's strategies to bring down the standard of living in Fria, Rusal was also confronted with aspirations in Débélé to reach purchasing power levels similar to those in Fria. It is thereby not astonishing that it rather tried to bring down the ones in Fria. However, given that the company controlled most of the collectively used infrastructure and social services, this amounted to austerity measures of a mini state rather than simple cuts in salaries.

4.3 Two Crises of Chain Governance

A comparison of the crises of chain governance in the 1970s and the 2000s shows that only little has changed during these 30 years. The involved African governments still try to maximize their stakes by juggling between different aluminum production networks that overlap quite neatly with global political fronts (cf. Larrue 1997: 113f). After WWII, these fronts ran between the two superpowers on the one hand and between the old colonial power and the US as the new potential hegemon on the other. Pechiney's special position as the only remaining French outpost in Guinea—which practically turned its corporate representatives into French ambassadors—also existed on a higher scale: while the French government tried to consolidate a European (until the 1960s even a Eurafrican) economic space—for instance, by pushing for a stronger protection of its aluminum industry—its corporate spin-off Pechiney supplied the Soviet Union with new technologies, e.g., by constructing the facilities in Nikolaiev. Today, these skirmishes within the West are less discernible, not least because power relations are clear, as European companies have definitely lost the scramble over the African aluminum production network against American companies. At present, Russian and Chinese companies do not cooperate in African bauxite mines, although the fact that Rusal went public in Hong Kong is nonetheless another indicator of the relatively tight bonds between the aluminum companies and their countries of origin, whether they are privately-owned entities or not. This notorious front between East and West has thereby remained the most exploited "mine" of Guinean politics: Alpha Condé—who has both a revolutionary background and good ties with the Western establishment—quite actively exploits this source.

However, recurrent predictions that more competing development partners, who additionally lower their aid conditionalities in their competition for resources, would inevitably result in greater gains for the population of the receiver countries (cf. Magrin 2013: 362) do not seem to apply to Guinea. After WWII, France had concrete plans for erecting a fully integrated aviation industry in Guinea (see Pré 1951). Out of self-interest, the French empire had tried to establish regionally integrated chains of production in Africa for at least half a century. A 662 km railway between Conakry and Kankan was already finished in 1914, after 12 years of construction. In 1958, with the retreat of the colonizers and the subsequent boycott of Guinean cash crops, this railway had to be closed down again (Dux 2000). More than a 100 years later, a similar project—the so-called *Transguinean*—which could serve various new mining projects with incomparably higher rates of return, no longer seems viable.[2] Furthermore, the success of Pechiney to create possibilities of downstream small-scale economic activity in Edéa (Cameroon) strongly depended on the territorialized colonial system of rule, even if this territorialization was weak and the French empire only just succeeded in creating territorialized production chains when it was already on the brink of disintegration.[3] On the other hand, the numerous broken promises to create integrated industrial complexes, since the 1960s[4] show the meager effects of polycephalous systems of control: with a multiplicity of hegemons, each trying to hedge his or her own interests, the possibility for nationally or regionally integrated production chains—thus the opposite of export enclaves—is considerably low (cf. Diallo et al. 2011: 55–56).

At the same time, resistance against this unified political system (the French empire) was easier to organize. Trade unions were the spearheads of independence and had been financed by French metropolitan parties. Guinea's first president, Sékou Touré, only refused de Gaulle's Eurafricanist visions when France announced the dissolution of French West Africa as a territorial unit and claimed the control of Guinea's raw materials (cf. Der Spiegel 1960). Today, both comprehensive development plans and coherent resistance movements seem to be hampered by the progressive fragmentation of political and economic space.

The greatest difference between the Southern uprisings of the 1960s and 1970s and renewed "resource nationalism" in the 2000s seems to be the current lack of South–South cooperation. Neither the experiments with producer cartels, such as the International Bauxite Association, nor Touré's and Nkrumah's early attempts to form a united (West) Africa have been repeated during the last mining boom (cf. Uwechue 1991: 755). The 2000s have rather seen pragmatic, nationally centered strategies that have failed to call into question African borders and governance structures (Diallo 2014).[5] These policies have not prevented a further corporatization of governmental tasks, despite the fact that neither the companies nor the governmental representatives are in favor of this development. TNCs initially felt forced to increasingly take over tasks of the state, because these were considered central to production (e.g., swimming pools to attract foreign experts, cranes to remove derailed trains, etc.). With the advent of CSR, as a response to democratic pressure from the consumer countries, TNCs even started to issue invitations to tender with their own local taxes and feel forced to pay public officials to supervise their own rehabilitation measures (Interview with Executive GAC 1/02/2014). They continue to do this because national democratic pressure on the Guinean government seems to be lower than transnational public pressure for CSR. Ultimately, this also means that private, despotically structured firm decision spaces seem to be attributed more political legitimacy than the elected government of Guinea.

4.4 Possible Ways Forward

The most promising way to more independent African decision-making processes would be a steep rise in the number of tax payers and, thus, the number of workers. A formalization and stabilization of small-scale mining seems to be as promising in this respect as the promotion of cash crops (cf. Kappès-Grangé and Soudan 2012). The erection of an integrated aluminum industry has been aspired since the 1950s by African decision makers and has remained a promise of governments and aluminum majors since then. Besides ecological and social issues particularly in connection with the needed energy and the low labor intensity of this sector, a recent and fairly detailed World Bank report also questions again the economic validity such projects (Husband et al. 2009). Apart from that, it does not seem very likely that power relations between the African governments and the global aluminum oligopoly would tremendously change if the former were in possession of smelters.[6] In addition, the WTO ruling against Chinese (bauxite) export taxes in 2012 further reduced the chances of weak governments to enforce the processing of their raw materials (WTO 2013: 88; cf. Küblböck 2014).

A multiplication of the economic bases of decision-making by creating jobs in small-scale mining and agriculture would lead to more governmental accountability, which in turn would provide a basis for the harmonization of national legal frameworks and the actual legal practice. As long as this is not the case, legal frameworks have to be hedged with legal provisions from outside. Instead of promoting mostly voluntary CSR measures (see Aluminium Stewardship Initiative 2015; cf. Hütz-Adams et al. 2014: 42; Brinkmann 2004), this could be done by the inclusion of social and ecological standards into bilateral trade agreements (BITs). Since 2009, the European commission concludes these agreements on behalf of its member countries (Global Policy Forum Europe 2011: 3–7; Kodweiß 2011: 2). Many other instruments that aim at democratizing corporate power, in general, could also be applied for the aluminum industry. For instance, this includes the closing down of tax heavens, the tightening of anti-trust laws, globally sanctioned social and ecological standards (e.g., by concluding framework agreements or by incorporating the WTO into the UN system), sector and countrywide minimum and maximum[7] wages (for instance, as called for by the Asia Floor Wage Campaign), the ban of commodity speculation, or fair and sustainable public procurement. If anything, this study should have shown that a sustainable amelioration of living conditions in Africa necessitates the reversal of global developments, which also increasingly endanger the living standards of all those who profit from Southern conditions of production until now.

Notes

1. This ratio is at the same scale as the US "CEO-to-worker compensation ratio", which stood at 510.7-to-1 in 2013. In 1978, the same ratio stood at 29.9-to-1 (Mishel and Davis 2014).
2. According to Dux, discussions on the *Transguinean* have already started in the 1970s (Dux 2000).
3. See Adamolekun (1976): 42–45, Amin (1973): XI, Condé (1972), Daviron and Ponte (2005): 85, Rempe (2011): 7, Touré (1967): 34.
4. Alcan in 1961 and 1974 (Campbell 2009: 79), Reynolds in 2002 (an unfulfilled extension project of Fria), Rusal in 2006 (ditto) and 2013 (Dian–Dian) and GAC only recently. According to insiders, the factory of GAC in Sangaredi will not be constructed in the next years to come, despite the resettlement of about 45 families.

5. The reactivation of the Mano River Union in 2004 is to date the only exception of this lack of regional co-operation. It had been formed in 1973 by Sierra Leone and Liberia and was joined by Guinea in 1980, but became inactive shortly thereafter.
6. One way to clarify this question would be a comparison with Suriname, which has produced aluminum out of its own bauxite since 1999 (cf. Brinkmann 2004: 22).
7. See e.g. Krugman (2014); Süddeutsche Zeitung (2013).

Bibliography

Adamolekun, Ladipo. 1976. *Sekou Toure's Guinea*. London: Methuen.

Aluminium Stewardship Initiative. (2015). ASI, http://aluminium-stewardship.org/, checked on 25/03/2015.

Amin, Samir. 1973. *Neo-Colonialism in West Africa*. Harmondsworth: Penguin.

Arrighi, Giovanni. 2002. The African Crisis. World Systemic and Regional Aspects. *New Left Review* 15: 5–36.

Bauer, Tobias, and Toya Maissen. 1989. Silbersonne am Horizont: ALUSUISSE—eine Schweizer Kolonialgeschichte. Zürich: Limmat-Verlag.

Brinkmann, Manfred. (2004). Silber aus Lehm. Wie nachhaltig ist Aluminium? Aachen: Brot für die Welt.

Campbell, Bonnie K. 2009. Guinea and Bauxite-Aluminium: The Challenges of Development and Poverty. In *Mining in Africa. Regulation & Development*, ed. Bonnie Campbell, 66–119. International Development Research Centre: Ottawa.

Châtelot, Christophe. (2015). Guinée: Dadis Camara candidat à la présidentielle. In *LeMonde.fr*, 12/05/2015.

Committee on foreign affairs. (2007). Prospects for Peace in Guinea. Hearing before the Subcommittee on Africa and Global Health of the Committee on Foreign Affaires. Serial No. 110–132. Washington, D.C: US Government Printing Office.

Condé, Alpha. 1972. *Guinée: Albanie d'Afrique ou néo-colonie américaine?*. Paris: Editions Git le Coeur.

Daviron, Benoit, and Stefano Ponte. 2005. *The Coffee Paradox. Global Markets Commodity Trade and the Elusive Promise of Development*. London: Zed Books.

Der Spiegel. (1960). Der Elefant. In spiegel.de, 16/03/1960, pp. 15–23.

Diallo, Ibrahima Talibé. (2014). Suite d'un bon dimanche to Johannes Knierzinger, 21/02/2014. E-Mail.

Diallo, Mamadou Chérif, et al. 2011. *Les enjeux de la gouvernance du secteur minier en Guinée*. Conakry: Cooperation Guinée Allemagne.

Dux, Jörg. (2000). Guinea—Raumpotential und wirtschaftliche Entwicklungsmöglichkeiten. Dissertation. Hamburg: Universität Hamburg.

Gereffi, Gary. 1997. Global Production Systems and third world development. In *Global Change, Regional Response: The New International Context of Development*, ed. Barbara Stallings, 100–143. Cambridge: Cambridge University Press.

Husband, Charles; McMahon, Gary; van der Veen, Peter (2009): The Aluminum Industry in West and Central Africa. Lessons Learned and Prospects for the Future. Extractive Industries and Development Series#13. Washington, D.C: World Bank.

Hütz-Adams, Friedel, Ann-Kathrin Voge, Sascha Liese, Stefan Hörmann, and Jan Per Bethge. 2014. *Nachhaltige Rohstoffe für den deutschen Automobilsektor. Herausforderungen und Lösungswege.* Bonn: Südwind e.V.

Investguinée (2012): Le patron de Rusal Oleg Déripaska. In investguinee.com, 14/04/2012, http://www.investguinee.com/portrait/le-patron-de-rusal-oleg-deripaska/47, checked on 12/08/2014.

Johnson, R.W. 1970. Sékou Touré and the Guinean Revolution. *African Affaires* 69 (277): 350–365.

Kappès-Grangé, Anne, François Soudan. 2012. Alpha Condé: "J'ai hérité d'un pays sans État". In Jeuneafrique.com, 16/07/2012, http://www.jeuneafrique.com/Article/JA2687p024-030.xml0/, checked on 9/08/2014.

Knierzinger, Johannes. (2014). Überlegungen zu Steuerung und ungleichem Tausch in Rohstoffgüterketten mit Beispielen aus Guineas Bauxitbergbau. In *Journal für Entwicklungspolitik* 30 (3).

Knierzinger, Johannes. (2016). Bauxit und Aluminium aus Afrika: Ausbeutung auf Umwegen. In Globale Ressourcen und Rohstoffpolitik. Aktuelle Auseinandersetzungen im historischen Kontext. Historische Sozialkunde/ Internationale Entwicklung 35, ed. Karin Fischer, Johannes Jäger, Lukas Schmidt, 203–220. Wien: New Academic Press.

Kodweiß, Marieke. 2011. *Unfairer Wettlauf um Rohstoffe. Die EU-Handels- und Rohstoofpolitik geht auf Kosten der armen Länder.* Berlin: Oxfam Deutschland/Weed.

Krugman, Paul. (2014). Why We're in a New Gilded Age. Review of "Capital in the Twenty-First Century". In The New York Review of Books, 8/05/2014, http://www.nybooks.com/articles/archives/2014/may/08/thomas-piketty-new-gilded-age/, checked on 11/07/2014.

Küblböck, Karin. (2014). Can 'Undistorted Access' Lead to Inclusive Development? The EU Raw Materials Initiative and Possible Effects upon Resource-Based Development in Africa. In *Journal für Entwicklungspolitik* 30 (3).

Larmer, Miles. 2006. *Mineworkers in Zambia: Labour And Political Change in Post-Colonial Africa.* London: Tauris.

Larrue, Jacques. 1997. *Fria en Guinée.* Karthala: Des aspects humains d'une industrialisation différente. Paris.

Lee, Ching Kwan. 2009. Raw Encounters: Chinese Managers, African Workers, and the Politics of Casualization in Africa's Chinese Enclaves. *The China Quarterly* 199: 647–666.

Magrin, Géraud. 2013. *Voyage en Afrique rentière*. Paris: Publications de la Sorbonne.

Mishel, Lawrence, Alyssa Davis. (2014). CEO Pay Continues to Rise as Typical Workers are Paid Less. Issue Brief#380. Washington, D.C.: Economic Policy Institute, http://s1.epi.org/files/2014/ceo-pay-continues-to-rise.pdf, checked on 13/06/2014.

Nyang'oro, Julius E. 1989. The State of Politics in Africa: The Corporatist Factor. *Studies in Comparative International Development* 24 (1): 5–19.

Pré, Roland. 1951. *L'avenir de la Guinée Française*. Conakry: Editions guinéennes.

Raikes, Philip, Michael Friis Jensen, Stefano Ponte. (2000). Global Commodity Chain Analysis and the French Filière Approach: Comparison and Critique. Copenhagen: CDR Working Paper.

Rempe, Martin (2011). Decolonization by Europeanization? The Early EEC and the Transformation of French—African Relations. KFG Working Paper No. 27. Berlin: Freie Universität Berlin.

Sassen, Saskia. 2005. The Global City. Introducing a Concept. *Brown Journal of World Affairs* 9 (2): 27–43.

Schmitter, Philippe C., and Gerhard Lehmbruch (eds.). 1979. *Trends Towards Corporatist Intermediation*. Beverly Hills: SAGE.

Shafer, D. Michael. 1986. Undermined: The Implications of Mineral Export Dependence for State Formation in Africa. *Third World Quarterly* 8 (3): 916–952.

Süddeutsche Zeitung. (2013). Nach Abzocker-Initiative: EU will Managergehälter begrenzen. In sueddeutsche.de, 4/03/2013, http://www.sueddeutsche.de/wirtschaft/nach-abzocker-initiative-in-der-schweiz-eu-kommission-will-managergehaelter-begrenzen-1.1615424, checked on 4/07/2014.

Taylor, Peter J. 2000. World Cities and Territorial States Under Conditions of Contemporary Globalization. The 1999 Annual Political Geography Lecture. *Political Geography* 19: 5–32.

Touré, Ahmed Sékou. 1967. *L'Afrique et la révolution*. Paris: Présence Africaine.

Uwechue, R. 1991. *Makers of Modern Africa*. London: Africa Books.

World Bank. (2014). World Development Indicators, http://data.worldbank.org/data-catalog/world-development-indicators, checked on 1/10/2014.

WTO. (2013). Dispute Settlement. Annual Report 2012. Geneva, http://www.wto.org/english/res_e/booksp_e/anrep_e/anrep13_chap5_e.pdf, updated on 31/05/2013, checked on 20/02/2015.

Index

J. Knierzinger, *Bauxite Mining in Africa*, International Political Economy Series, DOI 10.1007/978-3-319-52706-2

R

S

The manufacturer's authorised representative in the EU is Springer Nature Customer Service Centre GmbH, Europaplatz 3, 69115 Heidelberg, Germany. If you have any concerns regarding our products, please contact ProductSafety@springernature.com

Printed and bound by CPI Group (UK) Ltd, Croydon, CR0 4YY
25/09/2026
02222890-0005